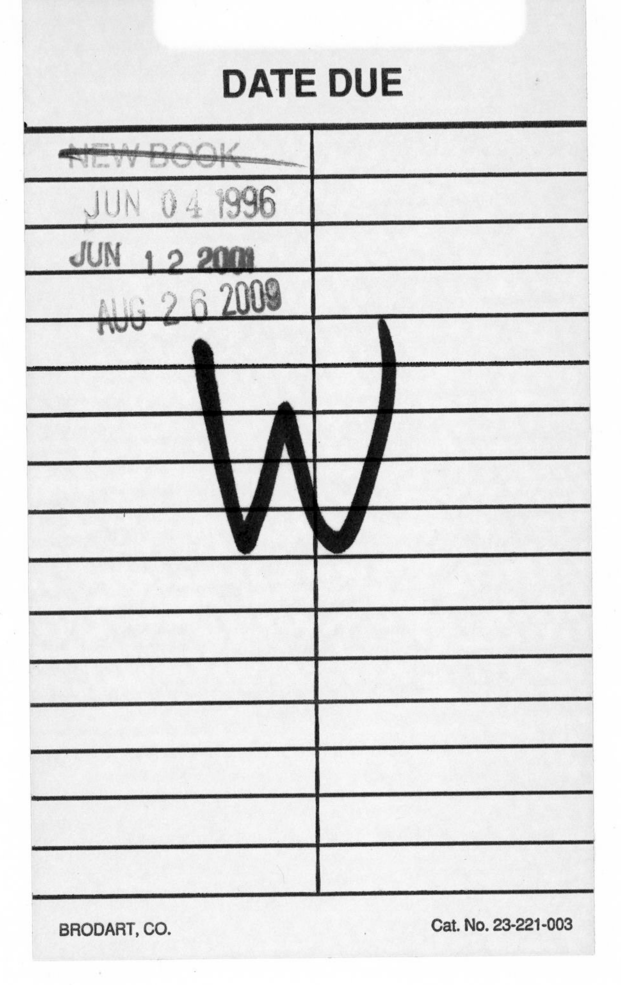
DATE DUE
NEW BOOK
JUN 04 1996
JUN 12 2001
AUG 26 2009
W
BRODART, CO.
Cat. No. 23-221-003

The Six Core Theories of Modern Physics

The Six Core Theories of Modern Physics

Charles F. Stevens

A Bradford Book
The MIT Press
Cambridge, Massachusetts
London, England

This book was set in Times Roman by Asco Trade Typesetting Ltd., Hong Kong, and was printed and bound in the United States of America.

First printing, 1995.

Library of Congress Cataloging-in-Publication Data

Stevens, Charles F., 1934–
The six core theories of modern physics / Charles F. Stevens.
p. cm.
"A Bradford book."
Includes bibliographical references and index.
ISBN 0-262-19359-0
1. Physics. 2. Mathematical physics. I. Title.
QC21.2.S688 1995
530.1—dc20 94-28453
CIP

Contents

Preface

As the title indicates, this book presents a brief, self-contained summary of the subjects that constitute the core of theoretical physics. The level is approximately that of advanced undergraduate or first-year graduate courses. In most cases, every step of the mathematics has been displayed (although I recognize that what is a step at one time can be a leap at another). Each chapter, except where noted with cross references, can be read independently of the others. Thus the book is intended to be used in any order for the six areas of physics covered.

I have in mind two audiences. The first comprises students of physics—advanced undergraduates or beginning graduate students—who are learning the material at this level for the first time. The book by itself would not be the place to start learning one of the six subject areas, because many important examples, applications, and extensions of the theories are not included and the presentation is too condensed to be the only source on any of the subjects. But the book should be a useful supplement to the standard texts. It offers an uncluttered overview of each topic in very few pages, often from a different vantage point than that of longer books, and provides the relevant material needed from related subject areas. For example, a succinct review of classical mechanics is provided to the student of quantum mechanics. The book should also be useful to the student who needs to review an area of physics rapidly.

The second audience consists of those in the quantitative sciences—engineering, biophysics, computer science, applied mathematics, economics, chemistry—who have known an area of physics once but have become rusty and want to relearn it. Often these individuals do not have the time or inclination to read an entire book but rather just need to brush up on the essentials of, say, electricity and magnetism. I anticipate that my brief treatment, with the required background material that I simultaneously provide (a summary of vector analysis, for example), will be useful to these readers.

The book summarizes most of the required mathematics and presents the basic theoretical structures of classical mechanics, electricity and magnetism, quantum mechanics, statistical physics, special relativity, and quantum field theory. Given the presumed level of sophistication of the audience, I tried to include everything needed to understand the material presented. The specific mathematical prerequisites differ somewhat from chapter to chapter but overall include some probability theory, calculus, differential equations, a familiarity with Laplace and Fourier transforms, and, in the last chapter, a bit of complex-variable theory. The first chapter summarizes most parts of applied mathematics used in this book. This presentation is not, however, intended to constitute an introduction to the areas of mathematics covered. For example, the student who has never studied linear algebra would find the dozen pages dedicated to a summary of linear operators on vector spaces too compact as an initial introduction to the subject. Someone who once knew about linear algebra should be able to relearn it fairly rapidly from what is presented. Specifically, the topics reviewed in more or less detail are vector analysis, linear operators on vector spaces, Green's functions for operator equations, the calculus of variations, the derivation of the diffusion equation, and tensor analysis. Functional techniques are used in four of the six main chapters, so the concept of a functional—together with functional differentiation, integration, and power series (Volterra expansion)—are also presented in the first chapter. As an example of the application of functional techniques, Gaussian random processes are presented from this point of view.

The main part of this book is contained in the six chapters on classical mechanics through quantum field theory. These chapters are designed to be read in any order, but the nature of the subject often means that knowledge from one area of physics is needed to understand another. The order in which the chapters appear indicates roughly their level of difficulty and the extent to which a particular chapter depends on knowledge and sophistication gained in preceding ones. "Classical Mechanics" stands alone and is required for most of the following chapters: "Quantum Mechanics," "Statistical Physics," and part of "Special Relativity" all draw on material presented in chapter 2, but the main sections of these chapters are independent of one another. "Electricity and Magnetism" is the chapter that is least directly connected to the other chapters; it depends on no other chapter and is used only in parts of later chapters that can be skipped. Throughout I have included cross references to the prerequisite earlier material so that in principle, if not in practice, the reader could start with the last chapter and understand it by referring to the cross-referenced sections of the earlier chapters.

Each chapter consists of a main part and additional optional sections indicated by the asterisk on the section heading. The main part gives the core theory, and the additional parts of the chapter present material that is necessary for something that comes later or that is more advanced and specialized. I recommend that these special sections be skipped unless the reader has some specific reason to include them. I have tried to avoid "it can be easily shown" steps in my development, although I know that in a few places this has crept in, particularly in parts of the first chapter, which is intended as a very condensed summary, and in the more specialized and advanced sections at the ends of chapters. If the manipulations necessary to progress from the start to the end of an equation chain do not quickly become clear, the reader should just continue on. I have often adopted the policy of having a result precede a detailed explanation of how the result was reached. My idea has been to let the reader see the goal before going through all of the steps needed to reach the goal. So one must develop the habit of looking ahead before spending a lot of time trying to figure out what manipulations were carried out. No equations are numbered, but the important equations have been given names, which are listed in the equation index. The point of excluding equation numbers is to eliminate such difficult-to-follow phrasing as, "Combining equation (3.17) with equation (9.117), we have ...." I have simply repeated earlier equations that do not appear on the same page but are needed and not named.

The "Classical Mechanics" chapter starts with the familiar second law of Newton and then reformulates it as the Euler-Lagrange equation, Hamilton's principle, and Hamilton's equations. The special section treats Poisson brackets, needed for the chapters on statistical physics and quantum mechanics.

"Electricity and Magnetism" begins with electrostatics and magnetostatics and then progresses to Maxwell's equations. This progression is followed twice, once to get Maxwell's equations for a vacuum and a second time to derive the macroscopic Maxwell's equations for fields in material media. The additional sections treat gauge transformations, the wave nature of the electromagnetic field in the absence of sources, and the Lorentz force on moving charges.

Quantum mechanics is treated, *à la* Feynman, with functional integrals. I adopted this approach both because it is a little more motivated and compact than the alternatives and because the functional-integral methods used in quantum mechanics now often appear in other areas of modern physics. The chapter follows Feynman and Hibbs to a great extent but includes more steps, different errors, and some material they did not cover (e.g., the uncertainty principle and the Heisenberg picture). The additional section of this chapter presents the formalism used to describe a driven harmonic oscillator and prepares one for the field theory treated in the last chapter.

I motivate my exposition of statistical physics with a quasi-historical treatment of thermodynamics and a faux-historical presentation of equilibrium statistical mechanics. The additional sections include an alternative presentation of equilibrium statistical mechanics, an extension to nonequilibrium processes, and a section on quantum statistical mechanics.

Special relativity starts with Einstein's two postulates and develops the equations usually needed for applying relativity to other areas of physics. The optional section presents the covariant form of Maxwell's equations.

The final chapter introduces modern field theories. Such a brief treatment cannot possibly cover an appreciable part of this topic in a comprehensible way, but I have tried to outline the logic of the theory and give some of the general results. My presentation is a natural extension of the quantum mechanics chapter to fields: it indicates how particles arise from fields, how particles interact with fields, how and why second quantization is used, and where antiparticles come from.

What is covered here has been derived from very many sources over a long time, and I can no longer remember the precise origin for most of the material I have presented. An annotated bibliography of books I find useful is provided as a guide to further reading for each of the chapters. As a graduate student, I studied with George Uhlenbeck and Mark Kac, and they have been the dominant influence on my approach to physics, as will be clear to those who knew them. Much of this book was written at the Aspen Center for Physics and the Santa Fe Institute, and I am indebted to these institutions for making their facilities available to me.

Notational Conventions

$\equiv$ means "is defined to be"

$\sim$ means "is proportional to"

$\partial_y f(x, y) \equiv \partial f(x, y)/\partial y$

$\partial_k f(\mathbf{x}) \equiv \partial f(\mathbf{x})/\partial x_k$, with $\mathbf{x} = (x_0, x_1, \ldots, x_N)$, an $(N + 1)$-dimensional vector

$\dot{y}(x) \equiv dy(x)/dx$, and $\ddot{y}(x) \equiv d^2y(x)/dx^2$

$\int f(x)\, dx$ generally denotes a definite integral with limits set by context

$\int f(x_1, \ldots, x_N)\, d^N x \equiv \underbrace{\int \cdots \int}_{N \text{ times}} dx_1, \ldots, dx_N\, f(x_1, \ldots, x_N)$

FOURIER TRANSFORM

$$\mathscr{F}[y(t)] \equiv \hat{y}(\omega) = \int_{-\infty}^{\infty} dt\, y(t) e^{-i\omega t} \qquad \text{(one-dimensional)}$$

$$\mathscr{F}[y(\mathbf{x})] \equiv \hat{y}(\mathbf{k}) = \int_{-\infty}^{\infty} d^N x\, y(\mathbf{x}) e^{-i\mathbf{k}\cdot\mathbf{x}} \qquad (N\text{-dimensional})$$

INVERSE FOURIER TRANSFORM

$$\mathscr{F}^{-1}[\hat{y}(\omega)] \equiv y(t) = \frac{1}{2\pi} \int_{-\infty}^{\infty} d\omega\, \hat{y}(\omega) e^{i\omega t} \qquad \text{(one-dimensional)}$$

$$\mathscr{F}^{-1}[\hat{y}(\mathbf{k})] \equiv y(\mathbf{x}) = \frac{1}{(2\pi)^N} \int_{-\infty}^{\infty} d^N k\, \hat{y}(\mathbf{k}) e^{i\mathbf{k}\cdot\mathbf{x}} \qquad (N\text{-dimensional})$$

The Six Core Theories of Modern Physics

1 Mathematics*

Although this chapter appears first, it should not be read first. Rather, the reader should start on any later chapter and then return to the relevant parts of this chapter as they are needed.

Vector Analysis

The purpose of this section is to record, rather than develop, some of the main results of vector analysis needed in the later chapters.

Algebra

A position in Cartesian space is specified by a radius vector:

$$\mathbf{r} = \mathbf{i}x + \mathbf{j}y + \mathbf{k}z,$$

where **i**, **j**, and **k** are the unit vectors along the x, y, and z axes, and x, y, and z are the components of the vector. A general vector **a** will also have three components, a_x in the x direction, etc. The magnitude of vector $|\mathbf{a}|$ (or $\|\mathbf{a}\|$) is defined to be

$$\|\mathbf{a}\| \equiv |\mathbf{a}| = \sqrt{a_x^2 + a_y^2 + a_z^2}.$$

The sum of two vectors is a new vector whose components are the sum of the corresponding components of the vectors being added.

Two kinds of vector multiplications are used: the *dot* (or *scalar*) product and the *cross* (or *vector*) product. The dot product of two vectors is a scalar:

$$\mathbf{a} \cdot \mathbf{b} = (a_x b_x) + (a_y b_y) + (a_z b_z) = |\mathbf{a}||\mathbf{b}| \cos\theta,$$

where θ is the angle between **a** and **b**.
The cross product of two vectors is another vector, which can be expressed as

$$\mathbf{a} \times \mathbf{b} = \mathbf{i}(a_y b_z - a_z b_y) + \mathbf{j}(a_z b_x - a_x b_z) + \mathbf{k}(a_x b_y - a_y b_x),$$

and it has the property

$$\mathbf{a} \times \mathbf{b} = -\mathbf{b} \times \mathbf{a}.$$

The magnitude of the vector $\mathbf{a} \times \mathbf{b}$ is

$$|\mathbf{a} \times \mathbf{b}| = |\mathbf{a}||\mathbf{b}| \sin\theta,$$

where θ is the angle between **a** and **b**. The vector $\mathbf{a} \times \mathbf{b}$ is normal to the plane defined by **a** and **b**, and the vector points in the direction for which, sighting along the vector, the rotation of **a** to **b** would be the shortest. Note that the magnitude of the dot product is greatest when the two vectors being multiplied are parallel but that the cross product vanishes under the same condition.

Some multiple products are used:

$$\mathbf{a} \times (\mathbf{b} \times \mathbf{c}) = (\mathbf{a} \cdot \mathbf{c})\mathbf{b} - (\mathbf{a} \cdot \mathbf{b})\mathbf{c}$$

$$(\mathbf{a} \times \mathbf{b}) \cdot (\mathbf{c} \times \mathbf{d}) = (\mathbf{a} \cdot \mathbf{c})(\mathbf{b} \cdot \mathbf{d}) - (\mathbf{b} \cdot \mathbf{c})(\mathbf{a} \cdot \mathbf{d})$$

$$\mathbf{a} \cdot (\mathbf{b} \times \mathbf{c}) = \mathbf{b} \cdot (\mathbf{c} \times \mathbf{a})$$

Calculus

A *field* is a function that associates one or more numbers with each point in space. Two types of fields are distinguished: scalar and vector, depending on whether the field function assigns a scalar or a vector value to each spatial location. The gravitational potential of the earth is a scalar field, and the force of gravity is a vector field (all forces point towards the center of the earth).

Three distinct types of spatial derivative operations are required for use with fields. The gradient, denoted as "grad" or ∇, operates on a scalar field and turns it into a vector field. The divergence, denoted as "div" or $\nabla \cdot$ (note the dot), operates on a vector field and turns it into a scalar field. Finally, the rotation, denoted "rot," "curl," or $\nabla \times$, operates on a vector field and transforms it into another vector field. Each of these field operations uses an operator, called *nabla* or *del*, defined as

$$\nabla \equiv \mathbf{i}\frac{\partial}{\partial x} + \mathbf{j}\frac{\partial}{\partial y} + \mathbf{k}\frac{\partial}{\partial z}.$$

This operator is treated as a vector, so that if it operates on a scalar, a vector will result (just as any scalar times a vector is another vector), etc. The gradient is del acting on a scalar field. The divergence is the dot product of del and a vector field, and the rotation is the cross product of del and a vector field. We consider these derivative types in turn.

Gradient The gradient of a scalar field $U(\mathbf{r})$, ∇U, is a vector whose magnitude at $\mathbf{r}$ is the largest rate at which U changes around $\mathbf{r}$ and whose direction is along the line of this greatest rate of change. Specifically,

$$\nabla U(x, y, z) = \mathbf{i}\frac{\partial U}{\partial x} + \mathbf{j}\frac{\partial U}{\partial y} + \mathbf{k}\frac{\partial U}{\partial z}.$$

A vector field that results from the gradient operation is called *conservative* because, as a consequence of the way it arises,

$$\oint_C \nabla U \cdot d\mathbf{s} = 0.$$

This is a line integral (with line element $d\mathbf{s}$) around any closed path. Any vector field whose line integral vanishes for all such closed paths is conservative.

Divergence The divergence is defined by a limiting process in terms of an integral of the vector field $\mathbf{E}(x, y, z)$ over the surface of a spatial volume. Picture an incompressible fluid moving so that the magnitude of $\mathbf{E}(\mathbf{r})$ is the flow velocity at $\mathbf{r}$ and the direction of $\mathbf{E}$ is the direction in which the fluid at $\mathbf{r}$ is moving. Enclose point $\mathbf{r}$ in a small volume. Then a surface element $d\boldsymbol{\sigma}$ of this volume defines a vector with magnitude $d\sigma$ (the area of the surface element) and direction along the outward normal to that small bit of surface. The volume of the enclosed area is V. The divergence is defined as

$$\nabla \cdot \mathbf{E}(\mathbf{r}) = \lim_{V \to 0} \frac{1}{V} \int_S \mathbf{E} \cdot d\boldsymbol{\sigma},$$

where the integral is taken over the entire surface S of the region enclosing $\mathbf{r}$. The quantity $\mathbf{E} \cdot d\boldsymbol{\sigma}$ is the amount of fluid flowing out through the surface element $d\boldsymbol{\sigma}$, and so the surface integral is the net amount of fluid flowing out of the enclosed region. If no sources or sinks of fluid are enclosed, then the net amount of fluid flowing through the enclosing surface must be zero, and $\nabla \cdot \mathbf{E} = 0$. If sources or sinks are present at $\mathbf{r}$, then $\nabla \cdot \mathbf{E}(\mathbf{r})$ gives the source density at $\mathbf{r}$.

The preceding discussion defined the divergence of $\mathbf{E}$ as $\nabla \cdot \mathbf{E}$, but this use of the $\nabla \cdot$ operator has not yet been justified. By taking the volume enclosing $\mathbf{r}$ as a tiny cube whose sides are parallel to the axes with areas $dxdy$, etc., can indeed show the use of $\nabla \cdot$ here to be correct.

For a very small volume dV, the definition of divergence gives

$$\nabla \cdot \mathbf{E}\, dV = \int_S \mathbf{E} \cdot d\boldsymbol{\sigma}.$$

Now construct an arbitrary volume of such small volumes, and sum up all of the contributions to give *Gauss's theorem*:

GAUSS'S THEOREM

(DIVERGENCE THEOREM)

$$\int_V \nabla \cdot \mathbf{E}\, dV = \int_S \mathbf{E} \cdot d\boldsymbol{\sigma}$$

Here only the surface of the volume appears on the right because the interior surfaces of the volume elements appear twice and cancel.

Rotation Not all fields are conservative. The rotation, or curl, $\nabla \times \mathbf{E}$ characterizes how much the field $\mathbf{E}$ departs from being conservative. Curl is defined, like divergence, as the limit of an integral, but since $\nabla \times \mathbf{E}$ is itself a vector, it has both a magnitude and a direction. Enclose point $\mathbf{r}$, as before, and let ΔS_u be the surface element whose normal points in the direction u. Then the component of $(\nabla \times \mathbf{E})_u$ in direction u is

$$(\nabla \times \mathbf{E})_u = \lim_{\Delta S \to 0} \oint_{\Delta S_u} \mathbf{E} \cdot d\mathbf{s}.$$

Here the line integral (with line element $\mathbf{ds}$) is taken around the contour defining the surface ΔS_u. Of course, if $\mathbf{E}$ is conservative, each of these line integrals will vanish. Since ∇U is always conservative, curl grad vanishes for every U:

CURL GRAD VANISHES

$$\nabla \times \nabla U = 0$$

As for the divergence, explicit calculation shows that the use of the $\nabla \times$ operator is correct for the rotation. Specifically,

$$\nabla \times \mathbf{E} = \mathbf{i}(\partial_y E_z - \partial_z E_y) + \mathbf{j}(\partial_z E_x - \partial_x E_z) + \mathbf{k}(\partial_x E_y - \partial_y E_x).$$

Note that here I have used the notation ∂_x in place of $\partial/\partial x$, etc.

If a large surface is made up of small surface elements, and if the contributions of each $\nabla \times \mathbf{E} \cdot d\boldsymbol{\sigma}$ are added together, the result is "Stokes's theorem":

STOKES'S THEOREM

$$\oint_C \mathbf{E} \cdot d\mathbf{s} = \int_S \nabla \times \mathbf{E} \cdot d\boldsymbol{\sigma}$$

The line integral is taken around the border of the surface. Here, as with Gauss's theorem, the contributions of the interior contours cancel because each enters twice with opposite values.

An important fact is that div curl vanishes for any $\mathbf{E}$:

DIV CURL VANISHES

$$\nabla \cdot \nabla \times \mathbf{E} = 0$$

To see why this is so, consider (for example) the top and bottom hemispheres of a globe and calculate the volume integral of $\nabla \cdot \nabla \times \mathbf{E}$ over both hemispheres. Use Gauss's theorem to get a pair of surface integrals and then Stokes's theorem to get a pair of line integrals (around the equator) to give

$$\int_V \nabla \cdot \nabla \times \mathbf{E}\, dV = \oint_C \mathbf{E} \cdot d\mathbf{s} - \oint_C \mathbf{E} \cdot d\mathbf{s} = 0.$$

The two line integrals cancel because they traverse the same contour (the equator) in opposite directions.

Helmholtz's theorem*

According to Helmholtz's theorem, a vector field $\mathbf{A}$ can be specified completely by its div and curl. To see how this works, write $\mathbf{A}$ as a sum,

$$\mathbf{A} = \mathbf{B} + \mathbf{C},$$

and suppose that we can find $\mathbf{B}$ and $\mathbf{C}$ (from $\mathbf{A}$) such that $\nabla \cdot \mathbf{B} = 0$ and $\nabla \times \mathbf{C} = 0$. Then

$$\nabla \cdot \mathbf{A} = \nabla \cdot (\mathbf{B} + \mathbf{C}) = \nabla \cdot \mathbf{C},$$

and

$$\nabla \times \mathbf{A} = \nabla \times (\mathbf{B} + \mathbf{C}) = \nabla \times \mathbf{B}.$$

Thus if we know the curl and the div of $\mathbf{A}$, by solving the appropriate differential equations we get $\mathbf{B}$ and $\mathbf{C}$, and these determine $\mathbf{A}$. This will be clear when $\mathbf{B}$ and $\mathbf{C}$ are explicitly exhibited in terms of $\mathbf{A}$.

Write $\mathbf{B}$ and $\mathbf{C}$ in terms of two potentials $\mathbf{M}$ and V:

$$\mathbf{B} = \nabla \times \mathbf{M} \qquad \mathbf{C} = -\nabla V$$

Since div curl $\mathbf{M}$ and curl grad V both automatically vanish, this means that $\nabla \cdot \mathbf{B} = 0$ and $\nabla \times \mathbf{C} = 0$. So the job is reduced to calculating $\mathbf{M}$ and

V from **A**. Start by examining the div and curl of **A** to see how they look in terms of the potentials **M** and V:

$$\nabla \cdot \mathbf{A} = \nabla \cdot \mathbf{C} = -\nabla^2 V$$

$$\nabla \times \mathbf{A} = \nabla \times (\nabla \times \mathbf{M})$$

Remember here the vector formula

$$\nabla \times (\nabla \times \mathbf{M}) = \nabla(\nabla \cdot \mathbf{M}) - \nabla^2 \mathbf{M}.$$

Thus,

$$\nabla \times \mathbf{A} = \nabla(\nabla \cdot \mathbf{M}) - \nabla^2 \mathbf{M}.$$

But **M** is a vector potential and can be chosen so that its div vanishes (as will be shown below). So any **M** with the correct properties will do. For this V and **M**,

$$\nabla \cdot \mathbf{A} = \nabla \cdot \mathbf{C} = -\nabla^2 V,$$

$$\nabla \times \mathbf{A} = \nabla \times \mathbf{B} = -\nabla^2 \mathbf{M}.$$

Note that the $\nabla \times \mathbf{A}$ equation is really three equations, one for each component. For example, for the x component it is

$$-\nabla^2 \mathbf{M}_x = (\nabla \times \mathbf{A})_x.$$

The above equations for $\nabla^2 V$ and each of the three components of $\nabla^2 \mathbf{M}$ are like Poisson's equation with driving functions $\nabla \cdot \mathbf{A}$ and (each of the j components of) $(\nabla \times \mathbf{A})_j$. Their solution is

$$V(\mathbf{x}) = \frac{1}{4\pi} \int \frac{\nabla' \cdot \mathbf{A}(\mathbf{x}')}{|\mathbf{x} - \mathbf{x}'|} d^3x'$$

and the three equations (one for each component)

$$\mathbf{M}(\mathbf{x}) = \frac{1}{4\pi} \int \frac{\nabla' \times \mathbf{A}(\mathbf{x}')}{|\mathbf{x} - \mathbf{x}'|} d^3x'.$$

By defining V and **M** in this way, we exhibit a decomposition of **A** into $\mathbf{B} + \mathbf{C}$ with the appropriate properties indicated above. This means that if the div and curl of **A** are given, they can be considered differential equations for **B** and **C** and solved to give **A**. This decomposition of a vector field **A** is sometimes said to be into the *longitudinal* and *transverse* parts of the field. The longitudinal component is **C** (its curl vanishes), and the transverse part is **B** (its div vanishes).

To know a vector field, then, one must know its div and curl, which is why Maxwell's equations must include divs and curls for the vector fields **E** and **B**.

Linear Operators on Inner-Product Spaces

Vector spaces

Inner-product spaces, vector spaces with a dot product defined, are a direct generalization, usually to more than three dimensions, of the ordinary three-dimensional vector spaces treated on page 1. As before, we have vectors (**a**, **x**, **e**, etc., lowercase boldfaced letters) and scalars (a, x, e, λ, etc., lowercase letters, never boldfaced), which are combined by the usual rules described earlier.

Notions of central importance for a general discussion of vector spaces are *linear dependence* and its opposite, *linear independence*. If some set of nonzero vectors $\{\mathbf{x}_j\}$ satisfies the relationship

$$\sum_j a_j \mathbf{x}_j = 0$$

for at least two nonzero scalars a_j, the vectors are said to be linearly dependent. Otherwise, they are linearly independent. If the $\mathbf{x}_j$ are linearly dependent, this just means that one of the vectors can be expressed as a linear combination of other vectors, and if the $\mathbf{x}_j$ are linearly independent, no vector $\mathbf{x}_j$ can be expressed as a linear combination of others. The vectors **i**, **j**, and **k** in ordinary three-space are linearly independent, and any vector in the space is a scalar multiple of, or can be expressed as a linear combination of, these three unit vectors. The dimension N of a vector space is the maximum number of linearly independent vectors in the space.

If this set of N linearly independent vectors in the N-dimensional space is augmented by any other nonzero vector in the space, the $N + 1$ vectors are now linearly dependent (because N is the maximum number of linearly independent vectors), so the new vector can be expressed as some linear combination of the first N vectors. Thus the original N vectors form a *basis*, which means that any vector in the space can be written as a linear combination of the N linearly independent vectors. Any N linearly independent vectors can serve as a basis, so a vector space can have more than one basis. The spaces I will use are either finite-dimensional or infinite-dimensional with a countable number of basis vectors.

An *inner* (or *dot* or *scalar*) product is defined as before to be a mapping from vector pairs to scalars: $(\mathbf{x}, \mathbf{y}) = \mathbf{x} \cdot \mathbf{y} = a$, where a is a scalar. The inner product, however, will be a little more complicated than the one used earlier, because we must permit the scalars to be the complex numbers required for some important physical applications (e.g., quantum mechanics). To define the inner product, we need to pick a basis $\{\mathbf{e}_j\}$, and expand two vectors $\mathbf{x}$ and $\mathbf{y}$ in this basis:

$$\mathbf{x} = \sum x_j \mathbf{e}_j \quad \text{and} \quad \mathbf{y} = \sum y_j \mathbf{e}_j,$$

where x_j and y_j are the scalar weights, usually called *components*, needed to form the two vectors from a linear combination of the basis vectors. Now define the *scalar* (or *inner* or *dot*) *product*:

SCALAR OR INNER OR DOT PRODUCT

$$(\mathbf{x}, \mathbf{y}) \equiv \mathbf{x} \cdot \mathbf{y} \equiv \sum_j x_j^* y_j \quad \text{or} \quad (\mathbf{x}, \mathbf{y}) \equiv \int du\, x^*(u) y(u),$$

where x^* is the complex conjugate of x. The second definition is used in some cases for which the space is infinite-dimensional, with u playing the role of the index j. Here the integral, according to our usual convention, is a definite one over the entire appropriate range and is a multidimensional integral if the functions depend on more than one variable. This inner product is just like the dot product defined earlier (page 1), except that the left-hand vector is represented by the complex conjugates of its x components. If the scalars are restricted to real numbers, the present definition reduces, of course, to the one used before. And as before, we define the *vector length* as follows:

VECTOR LENGTH

$$\text{length of } \mathbf{x} \equiv \|\mathbf{x}\| \equiv \sqrt{\mathbf{x} \cdot \mathbf{x}}$$

Because the rule for forming inner products involves the complex conjugates of components of the left-hand vector, it will be convenient to define a *dual vector space* with respect to the space in which the vectors we have been considering (like $\mathbf{x}$) reside. Each vector in our starting space has a counterpart in the dual space constructed by replacing all of the components by their complex conjugates. For example, the vector

$$\mathbf{x} = \sum_j x_j \mathbf{e}_j$$

has the counterpart in the dual space

$$\mathbf{x}^* = \sum_j x_j^* \mathbf{e}_j.$$

Note that the vector in the dual space that corresponds to $\mathbf{x}$ is denoted $\mathbf{x}^*$ and that we have arranged things so that the same $\{\mathbf{e}_j\}$ forms a basis for a space and its dual. From space/dual-space point of view, the inner product is a rule for assigning the combination of a vector and its dual to a scalar quantity. When an inner product is written $(\mathbf{x}, \mathbf{y})$ or $\mathbf{x} \cdot \mathbf{y}$, the components of the left-hand member are always taken to be those found in the dual of the indicated vector. Although I make the distinction for reasons that will become clear later in the discussion, the vector space and its dual in this case actually contain the same vectors.

When a pair of nonzero vectors $\mathbf{x}$ and $\mathbf{y}$ have a value 0 for their inner product, that is, whenever $\mathbf{x} \cdot \mathbf{y} = 0$, the vectors are said to be *orthogonal.* Further, when the length of each of the vectors in a pair is 1 ($\|\mathbf{x}\| = 1$ and $\|\mathbf{y}\| = 1$) and the vectors are orthogonal, then the pair is said to be *orthonormal.* A set of basis vectors that is orthonormal (as are the familiar **i**, **j**, and **k**) is very useful and can always be constructed from an arbitrary basis by a procedure called *Grahm-Schmidt orthonormalization.*

To carry out the Grahm-Schmidt procedure, start with an arbitrary basis $\{\mathbf{a}_j\}$ of an N-dimensional space. Suppose, for simplicity, that the space is real; the same procedure works for complex vector spaces, but the notation becomes a little more complicated. The goal is to turn this basis $\{\mathbf{a}_j\}$ into an orthonormal basis $\{\mathbf{e}_j\}$. Start by setting $\mathbf{e}_1 = \mathbf{a}_1/\|\mathbf{a}_1\|$, so $\mathbf{e}_1$ has a length of 1. Now define a vector $\mathbf{h} = \mathbf{a}_2 - (\mathbf{a}_2, \mathbf{e}_1)\mathbf{e}_1$. The vector $\mathbf{h}$ is just $\mathbf{a}_2$ with the part of $\mathbf{a}_2$ that points in the $\mathbf{e}_1$ direction subtracted away. Look at

$$\mathbf{h} \cdot \mathbf{e}_1 = (\underbrace{\mathbf{a}_2 - (\mathbf{a}_2, \mathbf{e}_1)\mathbf{e}_1}_{\mathbf{h}}, \mathbf{e}_1) = (\mathbf{a}_2, \mathbf{e}_1) - (\mathbf{a}_2, \mathbf{e}_1)\underbrace{(\mathbf{e}_1, \mathbf{e}_1)}_{=1} = 0.$$

Thus $\mathbf{h}$, constructed this way, is orthogonal to $\mathbf{e}_1$. Put $\mathbf{e}_2 = \mathbf{h}/\|\mathbf{h}\|$; we now have two orthonormal vectors $\mathbf{e}_1$ and $\mathbf{e}_2$. Continue this process and construct, on the kth step, a new $\mathbf{h}$ defined as

$$\mathbf{h} = \mathbf{a}_k - \sum_{j=1}^{k-1} (\mathbf{a}_k, \mathbf{e}_j)\mathbf{e}_j.$$

Now look at $(\mathbf{h}, \mathbf{e}_n)$ for any of the $\mathbf{e}_n$ produced in the first $k-1$ steps:

$$\begin{aligned}(\mathbf{h}, \mathbf{e}_n) &= (\overbrace{\mathbf{a}_k - \sum_{j=1}^{k-1} (\mathbf{a}_k, \mathbf{e}_j)\mathbf{e}_j}^{\mathbf{h}}, \mathbf{e}_n)\\ &= (\mathbf{a}_k, \mathbf{e}_n) - \sum_{j=1}^{k-1} (\mathbf{a}_k, \mathbf{e}_j)(\mathbf{e}_j, \mathbf{e}_n)\\ &= (\mathbf{a}_k, \mathbf{e}_n) - (\mathbf{a}_k, \mathbf{e}_n) = 0\end{aligned}$$

Again, $\mathbf{h}$ is orthogonal to each of the $k-1$ orthonormal vectors $\mathbf{e}_n$ that have already been constructed. We put $\mathbf{e}_k = \mathbf{h}/\|\mathbf{h}\|$, so that its length is 1. When this procedure had been carried out over N steps, we will have generated N mutually orthogonal vectors $\mathbf{e}_j$, all of length 1. Since these vectors are linearly independent (they are mutually orthogonal), they form a basis for the space. This is the desired orthonormal basis.

Linear operators

A linear operator $\mathbf{M}$ on a vector space assigns each vector in the space to a new vector (in the same space) according to the rule

$$\mathbf{M}(a\mathbf{x} + b\mathbf{y}) = a\mathbf{M}\mathbf{x} + b\mathbf{M}\mathbf{y} = a\mathbf{x}' + b\mathbf{y}',$$

with $\mathbf{x}'$ and $\mathbf{y}'$ being the new vectors assigned to $\mathbf{x}$ and $\mathbf{y}$ by $\mathbf{M}$. Note the convention of using boldfaced capital letters to denote operators.

If $\mathbf{A}$ and $\mathbf{B}$ are two linear operators, $\mathbf{AB} = \mathbf{BA}$ might well not be true. Whenever $\mathbf{AB}$ does equal $\mathbf{BA}$, then $\mathbf{A}$ and $\mathbf{B}$ are said to *commute*. Noncommuting operators are easy to construct and are common. For example, let $\mathbf{X}$ be the operator that means "multiply by x," let d/dx be the usual derivative operator, and take the differentiable functions $f(x)$ as the vectors. Then $x\, df(x)/dx$ certainly is not generally the same as $d(xf(x))/dx$, so $\mathbf{X}$ and d/dx do not commute.

Consider an arbitrary linear operator $\mathbf{M}$. With each operator $\mathbf{M}$ we associate another operator defined by the relation $(\mathbf{x}, \mathbf{Mx}) = (\mathbf{M}^*\mathbf{x}, \mathbf{x})$ for every vector. $\mathbf{M}^*$ is the *adjoint* of $\mathbf{M}$. An interpretation of what $\mathbf{M}^*$means (and why it exists) will become clearer, as will the need for the space/dual-space distinction, when we later consider the matrix interpretation of operators on page 15.

Several special operators are important. $\mathbf{I}$ is the identity operator, which sends every vector in the space to itself ($\mathbf{Ix} = \mathbf{x}$). Another important operator is the projection $\mathbf{E}_j$, which, for any vector $\mathbf{x}$, gives $\mathbf{E}_j\mathbf{x} = x_j\mathbf{e}_j$, where $\mathbf{e}_j$ is the jth member of an orthonormal basis and x_j is the jth component of $\mathbf{x}$. Thus the projection operator projects any vector onto one of the basis vectors.

Some operators have a companion operator that undo what the first operator has done. For example, if $\mathbf{Bx} = \mathbf{x}'$, the "undoing" would be $\mathbf{B}^{-1}\mathbf{x}' = \mathbf{x}$. Such an operator is called an *inverse*. Not all operators have inverses; for example, the projection operator $\mathbf{E}_j$ has no inverse, because many different $\mathbf{x}$s would be sent to the same $\mathbf{E}_j\mathbf{x} = x_j\mathbf{e}_j$. In fact, every vector with the component x_j for the basis vector $\mathbf{e}_j$ would be sent to the same $x_j\mathbf{e}_j$, so there would be no way of telling which of the possible vectors were sent there, and the inverse could not be defined. If an operator has no inverse, it is called *singular*. If it does have an inverse, the operator is called *nonsingular*.

Two classes of operators are of special importance in later chapters: the unitary operators, and the Hermitian operators. *Unitary* operators rotate all of the vectors in the space without changing the length of any vectors. *Hermitian* operators deform the space by stretching it in the direction of one or more basis vectors $\mathbf{e}_j$ for some orthonormal coordinate system $\{\mathbf{e}_j\}$. The rotating and stretching operators, along with the projection and identity operators, are the most important ones for most applications. Let us consider the unitary and Hermitian operators in turn.

Unitary operators Any operator $\mathbf{U}$ with the property that $\|\mathbf{Ux}\| = \|\mathbf{x}\|$ for all vectors $\mathbf{x}$ in the space is unitary; $\mathbf{U}$ does not change the length of vectors. Unitary operators have the interesting property that their adjoint is also their inverse. To see this, note that because $(\mathbf{x}, \mathbf{Ay}) = (\mathbf{A}^*\mathbf{x}, \mathbf{y})$ for any operator (definition of adjoint), $(\mathbf{Ux}, \mathbf{Ux}) = (\mathbf{U}^*\mathbf{Ux}, \mathbf{x})$ for every vector $\mathbf{x}$ in the space. Thus, because $(\mathbf{Ux}, \mathbf{Ux}) = (\mathbf{x}, \mathbf{x})$ (by the definition of the the unitary operator as length-preserving), $(\mathbf{U}^*\mathbf{Ux}, \mathbf{x}) = (\mathbf{x}, \mathbf{x})$ for every $\mathbf{x}$. This means that $\mathbf{U}^*\mathbf{U}\mathbf{x} = \mathbf{x}$, so $\mathbf{U}^* = \mathbf{U}^{-1}$. Unitary operators also send one orthonormal coordinate system (set of basis vectors) $\{\mathbf{e}_j\}$ into another orthonormal set of basis vectors $\{\mathbf{e}'_j\}$. If $\mathbf{e}'_j$ is the effect of $\mathbf{U}$ on basis vector $\mathbf{e}_j$, that is, if $\mathbf{e}'_j = \mathbf{U}\mathbf{e}_j$, then the vectors $\{\mathbf{e}'_j\}$ are orthonormal because

$$(\mathbf{e}'_i, \mathbf{e}'_j) = (\mathbf{U}\mathbf{e}_i, \mathbf{U}\mathbf{e}_j) = (\mathbf{U}^*\mathbf{U}\mathbf{e}_i, \mathbf{e}_j) = (\mathbf{U}^{-1}\mathbf{U}\mathbf{e}_i, \mathbf{e}_j) = (\mathbf{e}_i, \mathbf{e}_j) = \delta_{ij},$$

where δ_{ij} is the *Kronecker delta*:

KRONECKER DELTA

$$\delta_{ij} = \begin{cases} 1 & \text{if } i = j \\ 0 & \text{if } i \neq j \end{cases}$$

The N new vectors are linearly independent because they are orthonormal, and so they form a new orthonormal basis. The unitary operators thus must rotate the original coordinate system if they do anything at all.

Since $\|\mathbf{U}\mathbf{e}_j\| = \|\mathbf{e}_j\|$, it must be that $\mathbf{U}\mathbf{e}_j = \lambda_j \mathbf{e}_j$ for some complex number λ_j with the property that $|\lambda_j| = \lambda_j^* \lambda_j = 1$. Whenever an operator sends a particular vector into a scalar multiple of itself, that vector is called an *eigenvector* of the operator, and the scalar multiplying factor is called an *eigenvalue* corresponding to the eigenvector. So $\mathbf{e}_j$ is an eigenvector of $\mathbf{U}$, and λ_j is the corresponding eigenvalue. We have just shown that vectors in an orthogonal basis of our space are eigenvectors for all unitary operators, and the corresponding eigenvalues are complex numbers of the form $e^{i\theta}$ for some number θ. Unitary operators that work in a real vector space, rather than in a complex one, are called *orthogonal.*

Hermitian operators The class of all operators $\mathbf{H}$ that just deform the vector space by stretching coordinates are the Hermitian operators. For any orthonormal coordinate system $\{\mathbf{e}_j\}$, all of the operators that send $\mathbf{e}_j$ into $\lambda_j \mathbf{e}_j$ for some real number λ_j are Hermitian. Note that some of the λ_j might be negative, so "stretching" in this case would invert the coordinate. For a particular vector $\mathbf{x} = \sum x_j \mathbf{e}_j$, then, an Hermitian operator $\mathbf{H}$ will have the effect

$$\mathbf{H}\mathbf{x} = \sum x_j \mathbf{H}\mathbf{e}_j = \sum_j x_j \lambda_j \mathbf{e}_j.$$

Note that here we have selected the basis vectors to be the "natural" ones for $\mathbf{H}$, that is, the ones aligned in the directions that $\mathbf{H}$ stretches.

Hermitian operators have the interesting property of being *self-adjoint* (and the converse is also true). That is, $\mathbf{H} = \mathbf{H}^*$. To see this, consider a pair of vectors $\mathbf{x} = \sum x_j \mathbf{e}_j$ and $\mathbf{y} = \sum y_j \mathbf{e}_j$, and an Hermitian operator $\mathbf{H}$ that stretches by the amounts $\{\lambda_j\}$ (remember that these numbers are real) in the direction of the orthonormal basis $\{\mathbf{e}_j\}$. Look at

$$(\mathbf{H}^*\mathbf{x}, \mathbf{y}) = (\mathbf{x}, \mathbf{H}\mathbf{y}) = (\mathbf{x}, \sum y_j \lambda_j \mathbf{e}_j) = \sum x_j^* \lambda_j y_j = (\mathbf{H}\mathbf{x}, \mathbf{y}),$$

where $\mathbf{H}\mathbf{x} = \sum x_j \lambda_j \mathbf{e}_j$. I have used here the fact that the components of the dual of $\mathbf{H}\mathbf{x}$, $(x_j \lambda_j)^* = x_j^* \lambda_j$, because λ_j is real.

Since an Hermitian operator is defined to stretch in the direction of some basis vectors, we have

$$\mathbf{H}\mathbf{e}_j = \lambda_j \mathbf{e}_j$$

for the right basis $\{\mathbf{e}_j\}$. This means that the eigenvectors of an Hermitian operator form an orthonormal basis for the space, and the eigenvalues λ_j specify how much the space is stretched along each of the corresponding axes $\mathbf{e}_j$. For the basis $\{\mathbf{e}_j\}$ consisting of eigenvectors of $\mathbf{H}$, then, the operator $\mathbf{H}$ can be decomposed into projection operators $\mathbf{E}_j$ that project onto the $\mathbf{e}_j$. This representation gives the *spectral theorem*:

SPECTRAL THEOREM

$$\mathbf{H} = \sum_j \lambda_j \mathbf{E}_j$$

This theorem can be seen to be true by permitting $\mathbf{H}$ and $\sum \lambda_j \mathbf{E}_j$ to act on the same arbitrary vector $\mathbf{x}$ and showing that they produce the same result:

$$\mathbf{H}\mathbf{x} = \sum_j x_j \lambda_j \mathbf{e}_j$$

$$\sum_j \lambda_j \mathbf{E}_j \mathbf{x} = \sum_j \lambda_j \sum_k x_k \mathbf{E}_j \mathbf{e}_k = \sum_j \lambda_j x_j \mathbf{e}_j,$$

because

$$\mathbf{E}_j \mathbf{e}_k = \begin{cases} \mathbf{e}_j & \text{if } j = k \\ 0 & \text{if } j \neq k \end{cases}.$$

The eigenvalues $\{\lambda_j\}$ form the *spectrum* of the operator $\mathbf{H}$. Proving the spectral theorem here has not amounted to much, because the definition of Hermitian operators included the theorem. Usually Hermitian operators are defined as those that are self-adjoint, and what is proved is that all self-adjoint operators can be decomposed in the way specified by the spectral theorem.

The spectral theorem is of the greatest possible utility. To see why, consider some function of an operator $f(\mathbf{H})$. Functions of operators have important applications in physics. To give this function meaning, we might have to use its power-series expansion. For example,

$$e^{\mathbf{H}} \equiv \sum_n \frac{\mathbf{H}^n}{n!}.$$

Because powers of $\mathbf{H}$ are well defined ($\mathbf{H}$ is just applied n times to get $\mathbf{H}^n$), such functions are well defined. Whenever we can derive the spectral decomposition $\mathbf{H} = \sum_j \lambda_j \mathbf{E}_j$ of an operator,

$$f(\mathbf{H}) = \sum_j f(\lambda_j)\mathbf{E}_j,$$

because

$$\mathbf{H}^n\mathbf{x} = \sum_j x_j \mathbf{H}^n \mathbf{e}_j = \sum_j x_j \lambda^n \mathbf{e}_j.$$

Because powers of $\mathbf{H}$ are very easy to compute, so is its inverse:

$$\mathbf{H}^{-1} = \sum_j \frac{1}{\lambda_j}\mathbf{E}_j$$

and

$$\mathbf{H}^{-1}\mathbf{x} = \sum_j \frac{1}{\lambda_j} x_j \mathbf{e}_j.$$

Of course, an Hermitian operator is simple not in just any coordinate system but only in the coordinate system defined by the orthonormal eigenvectors of the operator. Suppose that $\{\mathbf{e}'_j\}$ is an orthonormal basis but $\{\mathbf{e}_j\}$ is a different orthonormal basis defined by the Hermitian operator $\mathbf{H}$, that is, that $\{\mathbf{e}_j\}$ are eigenvectors of $\mathbf{H}$ but $\{\mathbf{e}'_j\}$ are not. Define the unitary operator $\mathbf{U}$ that sends $\{\mathbf{e}'_j\}$ into $\{\mathbf{e}_j\}$ as $\mathbf{U}\mathbf{e}'_j = \mathbf{e}_j$. Now if $\mathbf{y} = \mathbf{H}\mathbf{x}$ for the basis $\{\mathbf{e}'_j\}$, we can change to the basis $\{\mathbf{e}_j\}$, where $\mathbf{H}$ is simple, by considering the vectors $\mathbf{U}\mathbf{y}$ and $\mathbf{U}\mathbf{x}$. Then $\mathbf{U}\mathbf{y} = \mathbf{H}\mathbf{U}\mathbf{x}$ or $\mathbf{y} = \mathbf{U}^{-1}\mathbf{H}\mathbf{U}\mathbf{x}$.

The composite operator $\mathbf{U}^{-1}\mathbf{H}\mathbf{U}$ is said to be the *diagonal form* of $\mathbf{H}$. We shall presently see why.

Hermitian operators stretch the vector space on which they act. How much to they deform it? A function, called the determinant, is used to answer this question. The *determinant* of an operator $\mathbf{M}$, $\det\{\mathbf{M}\}$, is defined by the relation

$$\det\{\mathbf{M}\} \equiv |\mathbf{M}| \equiv \prod_j \lambda_j,$$

where λ_j are the eigenvalues of $\mathbf{M}$. The value of the determinant specifies the factor by which a unit volume of the space is magnified (or shrunken). For example, if all the eigenvalues of an Hermitian operator that acts along the $\mathbf{i}$, $\mathbf{j}$, and $\mathbf{k}$ of 3-space were 2, a unit volume of the space would be increased 8-fold by the application of the operator. A unitary operator would not change the volume of the space (although it might make it negative).

The matrix representation of linear operators

The preceding discussion has not been tied to any particular coordinate system. When one coordinate system is selected, however, vectors and linear operators can be given a simple and very useful representation. Suppose that we pick a particular orthonormal basis $\{\mathbf{e}_j\}$ for an N-dimensional inner-product space and make it our standard one. Any vector $\mathbf{x}$ is written

$$\mathbf{x} = \sum_k (\mathbf{x}, \mathbf{e}_k)\mathbf{e}_k = \sum_k x_k \mathbf{e}_k,$$

with components $x_k = (\mathbf{x}, \mathbf{e}_k)$. Then as long as we keep to our standard basis, $\mathbf{x}$ can just as well be expressed in terms of its components by themselves:

$$\mathbf{x} = \begin{bmatrix} x_1 \\ x_2 \\ \vdots \\ x_N \end{bmatrix}$$

This $N \times 1$ array of components is called a *column vector.*

A similar representation can be achieved for a linear operator $\mathbf{A}$. When $\mathbf{A}$ acts on one of the basis vectors $\mathbf{e}_j$, the result $\mathbf{A}\mathbf{e}_j$ is another vector in the space, and so it can be expressed in terms of our standard basis:

$$\mathbf{Ae}_j = \sum_k a_{jk}\mathbf{e}_k,$$

where the coefficients a_{jk} are those required for the kth basis vector $\mathbf{e}_k$ in order to express the vector $\mathbf{Ae}_j$ in components. Specifically, $a_{jk} = (\mathbf{e}_k, \mathbf{Ae}_j)$, as can bee seen from the inner-product relation

$$(\mathbf{e}_k, \mathbf{Ae}_j) = \left(\mathbf{e}_k, \sum_i a_{ji}\mathbf{e}_i\right) = a_{jk}(\mathbf{e}_k, \mathbf{e}_k) = a_{jk}.$$

Since we know the effect $\mathbf{A}$ has on every basis vector, we automatically know the effect on any vector at all. Take vector $\mathbf{x}$ and look at $\mathbf{Ax}$. The effect of $\mathbf{A}$ on $\mathbf{x}$ is

$$\mathbf{Ax} = \mathbf{A}\underbrace{\sum_{k=1}^{N} x_k\mathbf{e}_k}_{\mathbf{x}} = \sum_{k=1}^{N} x_k\mathbf{Ae}_k = \sum_{k=1}^{N} x_k \underbrace{\sum_j a_{kj}\mathbf{e}_j}_{\mathbf{Ae}_k} = \sum_j \underbrace{\sum_{k=1}^{N} x_k a_{jk}}_{y_j}\mathbf{e}_j.$$

The components y_j of the vector $\mathbf{y} \equiv \mathbf{Ax}$ are thus given by the relationship

$$y_j = \sum_{k=1}^{N} x_k a_{jk}.$$

This same procedure can be carried out for any vector in the space. Thus, once we have the two-dimensional array of values a_{jk}, the effect of the operator $\mathbf{A}$ is completely determined, and just as a vector is represented by its components, the operator can be represented by a square array: the matrix that corresponds to $\mathbf{A}$. This if often written as $\mathbf{A} = [a_{jk}]$, where $[a_{jk}]$ is the square array with j specifying the row number and k the column. The (j,k)th entry in the matrix that represents the operator $\mathbf{M}$ will be denoted by $(\mathbf{M})_{jk}$.

Above I noted that the identity operator and the projection operator are important. What is the matrix representation of these operators? The identity operator $\mathbf{I}$ sends every vector into itself, and this, of course, includes the basis vectors $\mathbf{e}_j$. For the general operator $\mathbf{A}$, the element a_{jk} is given by the inner product $(\mathbf{e}_k, \mathbf{Ae}_j)$. For the identity operator $\mathbf{I}$,

$$(\mathbf{I})_{jk} = (\mathbf{e}_k, \mathbf{Ie}_j) = (\mathbf{e}_k, \mathbf{e}_j) = \delta_{jk}.$$

Thus all of the entries in the $\mathbf{I}$ matrix are 0 except for the diagonal elements (where $j = k$), all of which are 1.

For the projection operator $\mathbf{E}_k$ (which sends one of our standard basis vectors $\mathbf{e}_k$ into itself and all of the other basis vectors into $\mathbf{0}$), we have

$$(\mathbf{E}_k)_{jk} = (\mathbf{e}_k, \mathbf{E}_k\mathbf{e}_j) = (\mathbf{e}_k, \delta_{jk}\mathbf{e}_j) = \begin{cases} 1 & \text{if } j = k \\ 0 & \text{if } j \neq k \end{cases}.$$

Thus we see that the projection matrix that corresponds to the operator $\mathbf{E}_k$ contains entries of 0 everywhere except that the kth diagonal element $(\mathbf{E}_k)_{kk} = 1$.

The multiplication of operators was seen to be important in the preceding section. What is the matrix version of this? To find out, we look at the effect of the two operators $\mathbf{A}$ and $\mathbf{B}$ with matrix entries $a_{ij} = (\mathbf{A})_{ij}$ and $b_{ij} = (\mathbf{B})_{ij}$. From the above, we know that

$$\mathbf{y} \equiv \mathbf{A}\mathbf{x} = \sum_j y_j \mathbf{e}_j,$$

with $y_j = \sum_{k=1}^N x_k a_{jk}$. Now let $\mathbf{B}$ operate on $\mathbf{y}$:

$$\mathbf{B}\mathbf{y} = \mathbf{B}\mathbf{A}\mathbf{x} = \sum_i \underbrace{\sum_j b_{ij} y_j}_{(\mathbf{B}\mathbf{y})_i} \mathbf{e}_i$$

This means that the components of vector $\mathbf{B}\mathbf{y}$ are

$$(\mathbf{B}\mathbf{A}\mathbf{x})_i = (\mathbf{B}\mathbf{y})_i = \sum_j b_{ij} y_j = \sum_j b_{ij} \underbrace{\sum_k a_{jk} x_k}_{y_j} = \sum_k x_k \underbrace{\sum_j b_{ij} a_{jk}}_{(\mathbf{B}\mathbf{A})_{ik}}.$$

Thus the matrix that corresponds to $\mathbf{B}\mathbf{A}$ has entries

$$(\mathbf{B}\mathbf{A})_{ik} = \sum_j b_{ij} a_{jk}.$$

Hence the (i, k)th entry in the matrix $(\mathbf{B}\mathbf{A})$ is found by adding up the products of the ith row of $\mathbf{B}$ and the kth column of $\mathbf{A}$.

Operators can also be added, and the matrix interpretation of $\mathbf{A} + \mathbf{B}$ is especially easy. Because vectors are added by simply summing the corresponding components (see page 1), this means that matrices are also added by summing their corresponding components: $(\mathbf{A} + \mathbf{B})_{ij} = (\mathbf{A})_{ij} + (\mathbf{B})_{ij}$. Similarly, because multiplication of a scalar and a vector just results in the product for each of the components, multiplication of a matrix by a scalar just scales each of the components: $(b\mathbf{A})_{ij} = b(\mathbf{A})_{ij}$.

Operators are represented by a square matrix, and vectors by a list of components. A simplification can be achieved by viewing vector $\mathbf{x}$ as a matrix that has N rows and 1 column (our present space has N dimensions, and thus N basis vectors). That is,

$$\mathbf{x} = \begin{bmatrix} x_1 \\ x_2 \\ \vdots \\ x_N \end{bmatrix}.$$

With this definition, multiplying a matrix and a vector can be viewed simply as matrix multiplication:

$$\mathbf{A}x = \sum_j a_{ij}x_j = \sum a_{ij}x_{j\,1}$$

Here the vector $\mathbf{x}$ is viewed as a matrix with N rows and just one column, so its entries are all of the form $x_{j\,1}$. The result of this operation is another column vector with N rows.

How are we to view the inner product within this matrix framework? The inner product $(\mathbf{x}, \mathbf{x})$ is given by

$$(\mathbf{x}, \mathbf{x}) = \sum_j x_j^* x_j = \sum_j x_{1\,j}^* x_{j\,1},$$

where the vector $\mathbf{x}$ has, as above, been treated as a $N \times 1$ matrix. A consistent interpretation emerges if the dual of $\mathbf{x}$ is taken to be a $1 \times N$ row vector with entries $x_{1\,j}^*$. That is, the dual of $\mathbf{x}$ is

$$\mathbf{x}^* = [x_1^*, x_2^*, \ldots, x_N^*].$$

This gives a picture of the dual space: for each vector $\mathbf{x}$ represented by a column vector (i.e., a $N \times 1$ matrix) in our inner-product space, we have its dual represented by a row vector. The entries in this row vector are the complex conjugates of the corresponding entries in the original vector when the vector space is complex. For a real vector space, like the ordinary three-dimensional space, the dual of a column vector is simply the row vector with the same entries.

To multiply a matrix $\mathbf{A}$ by a dual (row) vector $\mathbf{x}^*$ (the dual of $\mathbf{x}$), we must place the row vector in front. That is,

$$\mathbf{x}^*\mathbf{A} = \sum_i x_{1\,i}^* a_{ij}$$

permits us to follow the usual rule for matrix multiplication and produces another row vector.

Next we seek a matrix interpretation of Hermitian operators. The definition of the adjoint of an operator **A** is $(\mathbf{x}, \mathbf{Ax}) = (\mathbf{A}^*\mathbf{x}, \mathbf{x})$, so for a self-adjoint operator, $(\mathbf{x}, \mathbf{Ax}) = (\mathbf{Ax}, \mathbf{x})$. For the left side of this equality we have

$$(\mathbf{x}, \mathbf{Ax}) = \sum_i x_i^* \left(\sum_j a_{ij} x_j \right) = \sum_{i,j} a_{ij} x_i^* x_j,$$

and for the right side,

$$(\mathbf{Ax}, \mathbf{x}) = \sum_j \left(\sum_i a_{ji} x_i \right)^* x_j = \sum_{i,j} a_{ji}^* x_i^* x_j.$$

Thus, for an Hermitian operator **A**, the corresponding matrix with entries a_{ij} has the property that $a_{ij} = a_{ji}^*$. For a real vector space (rather than a complex one), an Hermitian matrix **S** is symmetric; that is, $(\mathbf{S})_{ij} = (\mathbf{S})_{ji}$.

If the basis vectors chosen as the standard for defining matrices and vectors happen to coincide with the eigenvectors of an Hermitian operator **H**, then (by the SPECTRAL THEOREM, the matrix representation of projection operators, and the rules for adding matrices and multiplying them by scalars), **H** is just a diagonal matrix with the eigenvalues along the diagonal. If the Hermitian matrix is not diagonal, it can always be diagonalized, that is, put in a diagonal form, by the appropriate coordinate transformation. One must find only the unitary matrix **U** that changes the standard basis vectors to the ones that are eigenvectors for **H**. Then, as shown on page 14, the matrix $\mathbf{H}' = \mathbf{U}^{-1}\mathbf{HU}$ will be diagonal.

Hilbert space

A Hilbert space is an infinite-dimensional inner-product space with the property that if a sequence of Hilbert space vectors $\mathbf{x}_n$ approaches **x**, then **x** is also in the space. For the present purposes I will not worry much about the exact requirements for what does and does not make a legitimate infinite-dimensional vector space but will simply give some examples of the vectors and operators encountered. The vectors are functions $f(x)$, $g(x)$ of the sort usually encountered in applied mathematics. The inner product is defined as

$$(f, g) = \int dx\, f^*(x) g(x),$$

where the integral is a definite one taken over the entire suitable range, usually $(-\infty, \infty)$. An example of an operator is

$$\mathbf{K}f = \int dx'\, K(x, x')f(x').$$

This operator will be Hermitian if

$$(f, \mathbf{K}g) = \int\int dx dx'\, f^*(x)K(x, x')g(x')$$

$$= \int\int dx dx'\, K(x', x)f^*(x)g(x') = (\mathbf{K}f, g),$$

which will be true if $K(x, x') = K(x'x)$.

Because sines and cosines are orthogonal (we permit x to range only from 0 to 2π for the example), that is, because the inner product can be defined by

$$(e^{ijx}, e^{ikx}) \equiv \int_0^{2\pi} dx\, e^{-ijx}e^{ikx} = 2\pi\delta_{jk}$$

for integers j and k, these functions can be used as basis vectors with the inner product defined as indicated. The basis $\{\mathbf{e}_k\}$ is just the vectors (functions) $\{e^{ikx}\}$. An arbitrary vector (function) $f(x)$ in the Hilbert space can then be represented as

$$f(x) = \frac{1}{2\pi} \sum_{k=-\infty}^{\infty} F_k e^{ikx},$$

with

$$F_k = \int_0^{2\pi} dx\, f(x)e^{-ikx} = (e^{ikx}, f(x)).$$

This is just Fourier's theorem.

Operators also can be thought of as infinite-dimensional matrices. Recall that to define the effect of a linear operator, we must know just the effect on the basis vectors. Suppose that $\mathbf{M}$ is a well-behaved linear operator that acts on the functions of our space. It can be like the integral operator $\mathbf{K}$ described above. Now look at the effect $\mathbf{M}$ has on the kth basis vector e^{ikx}, and expand the result in terms of the basis:

$$\mathbf{M}e^{ijx} = \frac{1}{2\pi} \sum_{k=-\infty}^{\infty} m_{jk} e^{ikx},$$

where

$$m_{jk} = \int_0^{2\pi} dx\,(\mathbf{M}e^{ijx})e^{-ikx} = (e^{ikx}, \mathbf{M}e^{ijx})$$

(see page 16). Thus the operator **M** can be represented by an infinite matrix with entries m_{jk}. The Heisenberg formulation of quantum mechanics uses such matrices, but they do not appear in my treatment of the subject (chapter 4).

When the interval over which functions are defined stretches from $-\infty$ to ∞, the relations above become Fourier transforms. That is, the basis becomes $\{e^{i\omega x}\}$, where ω is a continuous variable, and the representation of a vector f in terms of components (denoted here by $\hat{f}$, the Fourier transform of f) and basis vectors is the integral

$$f(x) = \frac{1}{2\pi} \int_{-\infty}^{\infty} d\omega\, \hat{f}(\omega) e^{i\omega x},$$

with the components of the vector given by the inner product of f with basis vectors:

$$\hat{f}(\omega) = \int_{-\infty}^{\infty} dx\, f(x) e^{-i\omega x}$$

These basis vectors $\{e^{i\omega x}\}$ are the eigenfunctions for familiar operators. For example, the operator d/dx has the basis vectors as its eigenfunctions and has $i\omega$ as the corresponding eigenvalues:

$$\frac{d}{dx} e^{i\omega x} = i\omega e^{i\omega x}$$

Also, multiplication of vectors by 1 is the identity operator **I**, because $1 \cdot f = f$ for all vectors in the space; that is, every function is an eigenvector of 1, and the eigenvalue is 1. The eigenvalues of the operator $(d/dx + 1)$ are thus $\lambda_\omega = i\omega + 1$. We can use the SPECTRAL THEOREM, then, (although perhaps without complete justification) to solve the differential equation

$$(d/dx + 1) f(x) = g(x).$$

The formal solution is just

$$f(x) = \frac{1}{d/dx + 1} g(x).$$

So if we know the representation of $1/(d/dx + 1)$, we would have a solution. But this is just what the spectral theorem does. If $\hat{g}(\omega)$ are the components of the function g above, then, just as we calculated the inverse of an operator on page 14,

$$f(x) = \left(\frac{1}{d/dx + 1}\right) \underbrace{\frac{1}{2\pi} \int d\omega\, \hat{g}(\omega) e^{i\omega x}}_{g(x)} = \frac{1}{2\pi} \int d\omega \frac{\hat{g}(\omega)}{i\omega + 1} e^{i\omega x},$$

where $\lambda_\omega^{-1} = 1/(i\omega + 1)$. These results are, of course, the standard ones for the solution of differential equations with Fourier transforms.

Green's Functions

Green's functions (alternative names are *impulse response*, *influence function*, and *system function*) are of great use in many areas of applied mathematics. Here I discuss two closely related cases in which Green's functions appear. My presentation is in terms of specific examples, but the generalization to other situations is immediate.

First context

Suppose that we are to solve a linear-operator equation. We choose a simple first-order linear differential equation as an example, but any of a large class of linear operators are dealt with in just the same way:

$$\frac{dy}{dt} + ay(t) = \left(\frac{d}{dt} + a\right) y(t) \equiv \mathbf{M} y(t) = f(t),$$

where $\mathbf{M}$ is the operator $(d/dt + a)$. The *forcing* function $f(t)$ is given, and we are to find the response $y(t)$, which is taken to be 0 before we apply the forcing function at the initial time (that is, $y(t) = 0$ for $t < 0$). If we Laplace-transform the equation

$$\mathscr{L}[y(t)] \equiv \int_0^\infty dt\, y(t) e^{-st} \equiv \hat{y}(s)$$

(and set $\mathscr{L}[f(t)] \equiv \hat{f}(s)$), we get

$$(s + a)\hat{y} = \hat{f} \quad \text{or} \quad \hat{y}(s) = \frac{\hat{f}(s)}{s + a} = \left(\frac{1}{s + a}\right)\hat{f}(s).$$

Thus the solution involves two terms: $\hat{f}(s)$, the Laplace transform of the forcing function, and $\mathscr{L}[\mathbf{M}^{-1}] = 1/(s + a)$, the Laplace transform of the inverse of the operator **M**, which is usually called the *transfer function*. We call the inverse Laplace transform of operator **M** the *Green's function* for **M**:

$$\mathscr{L}^{-1}[1/(s + a)] = G(t)$$

By the convolution theorem,

CONVOLUTION THEOREM

$$\mathscr{L}\left[\int_0^t G(t - t')f(t')\,dt'\right] = \mathscr{L}[G(t)]\mathscr{L}[f(t)],$$

the solution to the original equation is the superposition equation:

SUPERPOSITION EQUATION

$$y(t) = \int_0^t dt'\, G(t - t')f(t')$$

Thus the Green's function $G(t)$ contains all of the information about the operator **M**.

The Dirac delta function

Before we proceed with the discussion, we need to take a brief side-path to define the *Dirac delta function*:

(DIRAC) DELTA FUNCTION

$$\delta(t - t') = 0 \quad \text{whenever } t \neq t'$$

$$\int_a^b dt\, \delta(t - t') = 1 \quad \text{if } (a < t' < b)$$

The delta function clearly becomes infinite in a special way when its argument is 0. One way to think of the delta function is to consider it as the derivative of a step function:

$$\delta(t) = \frac{d\theta(t)}{dt},$$

where θ is the unit step function:

$$\theta(t) = \begin{cases} 1 & \text{if } t \geq 0 \\ 0 & \text{if } t < 0 \end{cases}.$$

Thus $d\theta/dt = 0$ except at $t = 0$, and the integral of $d\theta/dt$ is 1, as it should be according to the defining properties of a delta function.

Many important properties of the delta function follow from these definitions. For example, there is the *screening property*:

SCREENING PROPERTY

$$f(a) = \int_{-\infty}^{\infty} dt\, f(t)\delta(t-a)$$

This means, for example, that the Fourier transform of the delta function is

$$\int_{-\infty}^{\infty} dt\, \delta(t-a)e^{-i\omega t} = e^{-i\omega a},$$

and the Laplace transform of a delta function is

$$\int_{0}^{\infty} dt\, \delta(t-a)e^{-st} = e^{-sa}.$$

Note that the Laplace transform (and the Fourier transform) of $\delta(t)$, that is, the case for which $a = 0$, is just $\mathscr{L}[\delta(t)] = \mathscr{F}[\delta(t)] = 1$.

Green's function as the impulse response

We return now to the main path. Associate with the earlier operator equation $\mathbf{M}y(t) = f(t)$ the equation

$$\mathbf{M}G(t) = \delta(t),$$

in which the original driving function $f(t)$ has been replaced by $\delta(t)$. $G(t)$ here is called the *impulse response*, as it is the response to delta-function (impulse) driving. Because the inverse Laplace transform of $\mathbf{M}^{-1}$ is defined to be $G(t)$ and because the Laplace transform of $\delta(t)$ is just 1, the Green's function defined above is the solution of this equation. Specifically, for the Laplace transforms,

$$\left(\frac{d}{dt} + a\right)G(t) = \delta(t)$$

gives

$$\hat{G}(s) = \frac{1}{s + a},$$

so

$$G(t) = \mathcal{L}^{-1}\left[\frac{1}{s + a}\right].$$

This is just the same result for the Green's function as we obtained earlier, and the arguments just given are easily generalized to other operators. The Green's function associated with an operator is thus just the solution to the operator equation with $\delta(t)$ used as a driving function. Of course, once the Green's function is known, the response to any driving function is calculated by a convolution of the driving function with the Green's function according to the SUPERPOSITION EQUATION.

Second context

Now we consider a partial differential equation with operator $\mathbf{P}$. I use the diffusion equation as an example:

$$\frac{\partial y(x,t)}{\partial t} - \frac{1}{2}\frac{\partial^2 y(x,t)}{\partial x^2} = 0$$

Thus,

$$\mathbf{P} = \frac{\partial}{\partial t} - \frac{1}{2}\frac{\partial^2}{\partial x^2},$$

and the homogeneous equation is

$$\mathbf{P}y(x,t) = 0.$$

Suppose that $y(x,t)$ vanishes for negative time and that at $t = 0$ we are provided with a known function as the boundary condition: $y(x, 0) = h(x)$. We Fourier-transform the operator equation and its boundary conditions with respect to the spatial variable (we use the Fourier transform now because x may range from $-\infty$ to ∞) and denote the Fourier-transformed functions, which have a transform variable k in place of x, with a *tilde* (for example, $\mathscr{F}[y(x,t)] \equiv \tilde{y}(k,t)$):

$$\frac{d\tilde{y}}{dt} + \frac{k^2}{2}\tilde{y}(k,t) = 0$$

$$\tilde{y}(k,0) = \tilde{h}(k)$$

The first-order equation for $\tilde{y}$ can be easily solved, and it will have an initial condition $\tilde{h}(k)$. Specifically,

$$\tilde{y}(k,t) = \tilde{h}(k)e^{-k^2t/2},$$

so that when the inverse Fourier transform is carried out to find $y(x,t)$, the result (by the CONVOLUTION THEOREM) will be the convolution of $h(x)$ with the inverse Fourier transform of $e^{-k^2t/2}$.

If we had used the pair of equations

$$MG(x,t) = 0 \quad \text{and} \quad G(x,0) = \delta(x),$$

then $\tilde{G}(k,t)$ would have been just $e^{-k^2t/2}$. This is a different sort of Green's function, one where the "driving" involves the boundary conditions rather than the differential equation itself. But in this case too, the solution to the more general problem is found with a convolution of the Green's function and the boundary-condition driving,

$$y(x,t) = \int_{-\infty}^{\infty} G(x - x', t)h(x')\,dx',$$

as can be seen from the preceding analysis.

Incidentally, the inverse Fourier transform of $e^{-k^2t/2}$ is

$$\mathscr{F}^{-1}[e^{-k^2t/2}] = \frac{1}{\sqrt{t}}e^{-x^2/2t},$$

so this is the Green's function for the diffusion equation.

The two versions of the Green's function can be united. If we consider a differential equation that is homogeneous and first-order (in time),

$$\frac{dy(t)}{dt} + ay(t) = 0 \quad \text{with} \quad y(0) = b,$$

then the solution with Laplace transforms is (in the *hat* notation)

$$s\hat{y}(s) - y(0) + a\hat{y}(s) = 0 \quad \text{or} \quad \hat{y}(s) = \frac{b}{s+a}.$$

The $y(0)$ appears because $\mathscr{L}[dy/dt] = \hat{y}(s) - y(0)$. This is just the same solution that would have resulted if we had required that $y(0) = 0$ and used the driving function $b\delta(t)$. The preceding homogeneous equation and the boundary condition would then have been replaced with

$$\frac{dy}{dt} + ay(t) = b\delta(t) \quad \text{and} \quad y(0) = 0.$$

Thus initial conditions can be included with delta-function driving. If this same thing is done for the partial-differential equation, the Green's function is described by

$$\left(\frac{\partial}{\partial t} - \frac{1}{2}\frac{\partial^2}{\partial x^2}\right)G(x,t) = \delta(t)\delta(x) \quad \text{and} \quad G(x,0) = 0^-$$

instead of

$$\left(\frac{\partial}{\partial t} - \frac{1}{2}\frac{\partial^2}{\partial x^2}\right)G(x,t) = 0 \qquad \text{and} \quad G(x,0) = \delta(x).$$

Green's functions will be particularly important in my treatments of quantum mechanics and field theory.

The Calculus of Variations

The calculus of variations is an interesting area of mathematics that has played an important role in the development of physics since the early eighteenth century. This method has been used particularly in classical mechanics and, more recently, has become important in field theory.

A simple calculus-of-variations problem can be stated as follows. Imagine a function $x(t)$ that depends on time t and has a derivative $\dot{x} = dx/dt$. The goal is to find the function $x(t)$ that makes

$$S[x] = \int_{t_s}^{t_e} L(x, \dot{x}, t)\, dt$$

a minimum (or more accurately, an extremum) for some given function L. Examples of how such a problem can arise are the following: (1) Among all the functions that start at $x(0) = a$ and end at $x(T) = b$, what path $x(t)$ in the (x, t) plane is the shortest? $S[x(t)]$ here is the curve length, and the job is to find the $x(t)$ that makes S a minimum. (2) Of all closed curves in the (x, t) plane with a fixed perimeter length P, which curve encloses the largest area?

The physical situation specifies the function L above. For instance, in the first example, the length ds of an element of a curve in the (x, t) plane is

$$ds = \sqrt{dx^2 + dt^2} = \sqrt{1 + \dot{x}^2}\, dt.$$

So that $L(x, \dot{x}, t) = \sqrt{1 + \dot{x}^2}$, and the function S to be minimized (with the restriction that $x(0) = a$ and $x(T) = b$) is

$$S[x] = \int_0^T \sqrt{1 + \dot{x}^2}\, dt.$$

In this particular case, x did not appear as an argument of L, but in general both x and $\dot{x}$ are present, because the behavior of a function at a particular point is determined by both its value and slope at that point.

The calculus-of-variations problem is solved by turning it into a differential equation known as the *Euler-Lagrange equation.* This equation is found by considering a variation δS produced by perturbing the correct $x(t)$ (that is, the function that actually makes S a minimum) by a small but otherwise arbitrary differentiable function $\eta(t)$ that vanishes at $t = 0$ and $t = T$ (so $x(0) = a$ and $x(T) = b$):

$$\delta S \equiv \int_0^T [L(x + \eta, \dot{x} + \dot{\eta}, t) - L(x, \dot{x}, t)]\, dt$$

The goal, then, is to find an equation specifying the $x(t)$ that makes $\delta S = 0$.

First we make use of the assumption that η is arbitrarily small and expand L to first order in its perturbed arguments:

$$L(x + \eta, \dot{x} + \dot{\eta}, t) \approx L(x, \dot{x}, t) + \eta \partial_x L + \dot{\eta} \partial_{\dot{x}} L$$

This expansion gives

$$\delta S = \int_0^T [\eta \partial_x L + \dot{\eta} \partial_{\dot{x}} L]\, dt.$$

Now we use integration by parts to change $\dot{\eta}$ to $-\eta$ and move the d/dt from $\dot{\eta}$ to $\partial_{\dot{x}} L$ (remember that η vanishes at the boundaries):

$$\delta S = \int_0^T [\eta \partial_x L + \dot{\eta} \partial_{\dot{x}} L)]\, dt = \int_0^T \underbrace{\left[\partial_x L - \frac{d}{dt}(\partial_{\dot{x}} L) \right]}_{=0} \eta\, dt$$

Note that because of the integration by parts, the total-time derivative of L appears here. Since η is an arbitrary function, the indicated part of the integrand must be 0 for δS to vanish (the condition for an extremum). This gives the *Euler-Lagrange equation* for this simple case:

ONE-DIMENSIONAL
EULER-LAGRANGE EQUATION

$$\frac{\partial L}{\partial x} - \frac{d}{dt}\left(\frac{\partial L}{\partial \dot{x}}\right) = 0$$

We need two generalizations of the Euler-Lagrange equation, both for problems in higher-dimensional spaces. The first generalization is really easy. Suppose that instead of a single variable $x(t)$, we need N variables $x_k(t)$ to describe a system, where $k = 1, 2, \ldots, N$ and each x_k is a function of t. If each of these variables is independent of the others, then we can carry out the preceding procedure with a variation of each variable separately and come up with N Euler-Lagrange equations that are just like the one above except that x_k replaces x and $\dot{x}_k$ replaces $\dot{x}$.

The second generalization involves arguments of L that depend on $\mathbf{x}$, an N-dimensional vector. I call these arguments $\phi(\mathbf{x})$. Now to find S we must integrate over an N-dimensional space:

$$S[\phi] = \int_a^b L(\phi(\mathbf{x}), \partial_k \phi)\, d^N x$$

The same line of reasoning used before leads to

$$\begin{aligned}\delta S &= \int_a^b \left[\delta\phi \frac{\partial L}{\partial \phi} + \sum_k \delta(\partial_k \phi) \frac{\partial L}{\partial(\partial_k \phi)}\right] d^N x \\ &= \int_a^b \left[\frac{\partial L}{\partial \phi} - \sum_k \frac{\partial}{\partial x_k}\left(\frac{\partial L}{\partial(\partial_k \phi)}\right)\right] \delta\phi \, d^N x.\end{aligned}$$

Thus for this multidimensional case, we have the *general Euler-Lagrange equation*:

GENERAL EULER-LAGRANGE EQUATION

$$\frac{\partial L}{\partial \phi} - \sum_k \frac{\partial}{\partial x_k}\left(\frac{\partial L}{\partial(\partial_k \phi)}\right) = 0$$

Of course, ϕ might also be multidimensional, so that ϕ_j would replace ϕ in the preceding equation.

Random Walk

Einstein derived the diffusion equation from a random walk. Here I give a version of this derivation for the one-dimensional case because the resulting equation, and its interpretation, will be of importance later.

Picture a particle that moves at random in one dimension away from its initial location of $x = 0$. We divide time into ticks of length Δt, and the x axis, where the particle is moving, into discrete steps of length Δx. We suppose that the particle must take a step for each Δt and that it can step only to a neighboring location on the x axis. The probability of a right step = the probability of a left step = 1/2 and is independent of time and the particle's position. We have thus assumed that the particle moves according to a Markov process: the behavior in the next time increment depends only on the particle's current state (position, here) and not on the path that led to that position.

Let $p(x, t)$ be the probability that the particle is at position x at time t. A particle can arrive at this position only by being at an adjacent location, either $x - \Delta x$ or $x + \Delta x$, at the immediately preceding time $t - \Delta t$:

$$p(x, t) = \tfrac{1}{2} p(x - \Delta x, t - \Delta t) + \tfrac{1}{2} p(x + \Delta x, t - \Delta t)$$

Recall that

$$\Delta f(x) \equiv f(x) - f(x - \Delta x)$$

and that

$$\Delta^2 f(x + \Delta x) \equiv \Delta\Delta f(x + \Delta x)$$
$$= \Delta[f(x + \Delta x) - f(x)]$$
$$= f(x + \Delta x) - 2f(x) + f(x - \Delta x).$$

Since $p(x, t)$ depends on two variables, we must use Δ_x to denote operations on the x argument and Δ_t for operations on the t argument.

We subtract $p(x, t - \Delta t)$ from both sides of the equation to get

$$p(x, t) - p(x, t - \Delta t)$$
$$= \tfrac{1}{2}p(x - \Delta x, t - \Delta t) - p(x, t - \Delta t)$$
$$+ \tfrac{1}{2}p(x + \Delta x, t - \Delta t).$$

With the definition of the Δ operators, this last equation may be rewritten as

$$\Delta_t p(x, t) = \tfrac{1}{2}\Delta_x^2 p(x - \Delta x, t - \Delta t).$$

And after we divide by Δt, this becomes

$$\frac{\Delta_t p(x, t)}{\Delta t} = \left(\frac{\Delta x^2}{\Delta t}\right)\frac{1}{2}\,\frac{\Delta_x^2 p(x - \Delta x, t - \Delta t)}{\Delta x^2}.$$

We then take the limit as Δx and Δt approach 0 in such a way that $\Delta x^2/\Delta t$ approaches 1 in the units of the problem which gives the *diffusion equation*:

DIFFUSION EQUATION

$$\frac{\partial p(x, t)}{\partial t} = \frac{1}{2}\,\frac{\partial^2 p(x, t)}{\partial x^2}$$

For initial condition ξ and τ, this equation has the solution

$$p(x, t; \xi, \tau) = \frac{1}{\sqrt{2\pi(t - \tau)}}e^{-(x-\xi)^2/2(t-\tau)}$$

(see page 26). Here $p(x, t; \xi, \tau)$ is the conditional probability of finding the particle at position $x(t)$, given that it started at $\xi(\tau)$.

A random walk in continuous time of this sort is a *Wiener process*, and it plays a central role in probability theory.

Functional Calculus

The purpose of this section is to introduce the concept of a *functional* and to describe briefly the functional calculus. Functionals are used in a variety of areas of modern physics; for example, they play a central role in the Feynman treatment of quantum mechanics presented in chapter 4.

A functional takes a function as an argument, as does a function of a function, and assigns it to a number. However, a function of a function just looks at the value of the argument function, but a functional looks at the entire behavior of its argument function. Let us start with a specific example to make this distinction clear. If $f(x) = x^2$ and $g(x) = e^{-x}$, then

$$y \equiv g(f(x)) = e^{-f(x)} = e^{-x^2}$$

is a usual function of a function to which, for example, the chain rule could be applied for taking derivatives. For each value of x, we would then assign a value to y by evaluating $y = g(f(x))$ at x.

The functional $y = g[f(x)]$ (note that square brackets are used to make the distinction between a usual function of a function and a functional) would also have a value that depends on the function $f(x)$. The difference is that now y would depend not on the value of $f(x)$ at a particular point x, but on the entire path of $f(x)$ as x varies over some specified range (say from 0 to 1). If $f(x) = x^2$ as before, y might, for example, be the value of the functional $g[f]$ defined as

$$y = g[f(x)] \equiv \int_0^1 f(x)\,dx = \int_0^1 x^2\,dx.$$

Here y depends on the entire behavior of $f(x)$ over the range $x \in [0, 1]$. If we used a different argument for the functional, say $f(x) = \sqrt{x}$, then

$$y = g[f(x)] = \int_0^1 f(x)\,dx = \int_0^1 \sqrt{x}\,dx$$

would have a different value. Note that if g were defined as

$$g[f(x), w] = \int_0^w f(x)\,dx$$

(note the upper limit of the integration), g would be a functional of f and a function of w.

An example of the use of functionals appeared in the preceding section "The Calculus of Variations," where we found the function $x(t)$ that minimized the functional $S[x]$ defined there.

A continuous functional is defined like a continuous function: by perturbing the argument of the functional by a sufficiently small amount, we can ensure that the functional's value is changed by as little as we please. The example of the functional $y = g[f]$ given above is continuous because of the properties of the integral.

Functionals can be thought of as extensions of functions of many variables. Most of the functions with which we will be concerned can be well approximated by their values at N locations, and the approximation becomes better and better as N increases. If we take an array of N values $u_1, u_2, \ldots, u_k, \ldots, u_N$, where $u_k = f(x_k)$ for successive $x_1, x_2, \ldots, x_N$, then a sequence of functions $g_N(u_1, \ldots, u_N)$ can be found that approaches the functional $g[f(x)]$ as the spacing between the x_ks becomes closer and closer. For example, we can divide up the x axis from 0 to 1 in N equal steps with values $x_0 = 0, x_1 = \Delta x, \ldots, x_k = k\Delta x, \ldots$ and represent the function $f(x) = x^2$ considered above as the array $x_0^2, x_1^2, \ldots$. The functional $g[f(x)] = \int f\,dx$ used as an example above could be approximated for $f(x) = x^2$ by

$$g_N(x_0, x_1, \ldots, x_{N-1}) \equiv \sum_{k=0}^{N-1} x_k^2 \Delta x$$

$$\approx g[x^2] = \int_0^1 x^2\,dx.$$

As N increases, g_N differs less and less from $g[x^2]$, and in the limit of $N \to \infty$, this representation becomes exact, of course.

More generally, take a functional $F[\phi(x)]$. To define a series of functions of many variables that approach F, let us suppose that F is a continuous functional and require that its arguments ϕ also be continuous. We can construct a piecewise linear approximation to the argument $\phi(x)$ of the functional $F[\phi]$ with the continuous function $\hat{\phi}_N(x)$ defined so that $\hat{\phi}_N(x) = \phi(x_k) \equiv \phi_k$ for $x = x_0, \ldots, x = k\Delta x, \ldots, x = N\Delta x$, with $\hat{\phi}$ varying linearly between the ϕ_k to connect them continuously. Thus $\hat{\phi}_N(x) \to \phi(x)$ as $N \to \infty$. Further, we can define a function of N variables $F_N(\phi_1, \ldots, \phi_n) \equiv F[\hat{\phi}_N(x)]$ because $\hat{\phi}$ is completely specified by the N values of ϕ_k. Since F is continuous, $F_N \to F[\phi]$ as $N \to \infty$. Here, then, we can see explicitly how a large class of functionals can be represented as the limits of functions of N variables as N increases. This representation will be important in our later discussions.

In what follows, we define functional derivatives, functional power series, and functional integration. Each of these will turn out to be the natural analog of the corresponding derivative, power series, and integral of a function of many variables.

Functional derivatives

Functional derivatives can be defined in two different ways. The first definition is like the definition of an ordinary derivative:

FUNCTIONAL DERIVATIVE
(FIRST DEFINITION)

$$\frac{\delta F[\phi]}{\delta\phi(\xi)} = \lim_{h\to 0}\frac{1}{h}(F[\phi(x) + h\delta(x-\xi)] - F[\phi(x)])$$

Here the functional derivative is denoted by $\delta F[\phi]/\delta\phi(\xi)$. Note that the functional derivative is a function of ξ, the point at which the argument $\phi(x)$ is perturbed by the δ function.

For example, if $F[\phi(x)] = \int g(x)\phi(x)\,dx$, then

$$\begin{aligned}\frac{\delta F[\phi(x)]}{\delta\phi(\xi)} &= \lim_{h\to 0}\frac{1}{h}\left(\int g(x)(\phi(x) + h\delta(x-\xi))\,dx - F[\phi(x)]\right)\\ &= \lim_{h\to 0}\frac{1}{h}\int g(x)h\delta(x-\xi)\,dx\\ &= g(\xi).\end{aligned}$$

A related example of this definition, and an especially simple one, uses the degeneration functional $F[\phi(x)] = \phi(x) = \int\phi(\eta)\delta(x-\eta)\,d\eta$:

$$\begin{aligned}\frac{\delta F[\phi]}{\delta\phi(\xi)} &= \lim_{h\to 0}\frac{1}{h}(\phi(x) + h\delta(x-\xi) - \phi(x))\\ &= \delta(x-\xi)\end{aligned}$$

This is a degenerate functional—just a function, really—because its value depends on the value of its argument at one point $\mathbf{x}$ rather than on an entire function.

The functional derivatives used so far have been one-dimensional; that is, their argument function ϕ has depended on a single variable $\mathbf{x}$. In general, of course, the argument function might depend on a vector $\mathbf{x}$. For this situation, the functional derivative is defined to be

$$\frac{\delta F[\phi(\mathbf{x})]}{\delta\phi(\mathbf{x}')} = \lim_{h\to 0}\frac{1}{h}(F[\phi(\mathbf{x} + h\delta(\mathbf{x} - \mathbf{x}')] - F[\phi(\mathbf{x})]),$$

where $\delta(\mathbf{x} - \mathbf{x}') = \prod_j \delta(x_j - x_j')$, with x_j and x_j' being components of the vectors $\mathbf{x}$ and $\mathbf{x}'$.

We could use some function other than the delta function to perturb ϕ as long as the perturbing function vanishes everywhere except near one location $\mathbf{x}$ and becomes arbitrarily small at that location as h approaches 0. For example, an appropriate Gauss curve with a variance proportional to h could be used as the perturbing function. This is really like using a delta function without identifying it as such. Suppose that perturbing function is called $\eta_h(x, \xi)$, where h is a parameter that makes the perturbation become smaller and smaller and more sharply localized around x as $h \to 0$. The functional derivative would then be defined as the limit

$$\frac{\delta F}{\delta\phi(x)} = \lim_{h\to 0}\frac{F[\phi(\xi) + \eta_h(x,\xi)] - F[\phi(\xi)]}{\int \eta_h(x,\xi)\,d\xi}.$$

The notion of a functional as the limit of a function of N variables, as described above, provides insight into the meaning of functional derivatives and also marks the path for developing their second definition. First I must stress that just as $\partial g(u_1,\ldots,u_N)/\partial u_k$ depends on its index k, so the functional derivative $\delta F[\phi]/\delta\phi(\xi)$ depends on its "index" ξ, that is, on the place at which the argument $\phi(x)$ is perturbed by the delta function. The convention is that the "index" is the argument for the function being perturbed. That is, if the perturbing function is localized around ξ, the index ξ appears as the argument of $\phi(\xi)$ in the expression $\delta F[\phi]/\delta\phi(\xi)$.

Second, note that the variation of the functional $\delta S[x]$ used in the preceding section "The Calculus of Variations" is the functional analog of the standard variation $dF_n(\phi_1, \phi_2, \ldots, \phi_N)$. For a function of N variables,

$$dF_N \approx \sum_k^N \frac{\partial F}{\partial \phi_k} d\phi_k.$$

If $F_N(\phi_1,\ldots,\phi_N) \to F[\phi(x)]$, then each time N is increased, the approximation generally becomes better. Because $dF_N \to \delta F[\phi]$, increasing N makes dF_N approach the fixed value $\delta F[\phi]$. This means that when N and the number of terms in the sum

$$dF_N = \sum_k^N \frac{\partial F}{\partial \phi_k} d\phi_k$$

are increased, the value of the sum approaches a fixed value. Thus the terms in the sum must correspondingly decrease. As $\phi(x)$ is continuous, the adjacent ϕ_k are hardly different if N is sufficiently large. On the other hand, $\partial F_N/\partial \phi_k$ must decrease in proportion to $1/N$ because the extent of the perturbation of $\phi(x)$ is decreased as the spacing between the values becomes less. For a large N, then, the term $(\partial F_N/\partial \phi)\, dx$ approaches the function $F'(x)\, dx$, and

$$\delta F[\phi(x)] \to \int_x F'(x)\eta(x)\, dx$$

for the arbitrary perturbing function $\eta(x) = \delta\phi(x)$. Because $F'(x)$ is the "density" of $\partial F_N/\partial \phi$, this function can be identified as the functional derivative $\delta F/\delta \phi$:

FUNCTIONAL DERIVATIVE
(SECOND DEFINITION)

$$\delta F[\phi] = \int_\xi F'(\xi)\eta(\xi)\, d\xi$$

The functional derivative is identified by the function

$$\frac{\delta F[\phi]}{\delta \phi(\xi)} = F'(\xi)$$

for an arbitrarily small function $\eta(\xi)$ that perturbs $\phi(x)$.

An example reveals the equivalence of these two definitions. Suppose that we define the functional F as

$$F[g(x), y] = \int_0^\infty K(x, y)g(x)\, dx$$

for some suitable kernel $K(x, y)$; F is a functional of g and a function of y. By the first definition, the functional derivative is then

$$\frac{\delta F[g,y]}{\delta g(\xi)} = \lim_{h\to 0}\frac{1}{h}\left(\int_0^\infty K(x,y)(g(x)+h\delta(x-\xi))\,dx - F[g,y]\right)$$

$$= \lim_{h\to 0}\frac{1}{h}\int_0^\infty K(x,y)h\delta(x-\xi))\,dx$$

$$= K(\xi,y).$$

Thus $\delta F/\delta g(\xi) = K(\xi, y)$. Note that the resulting functional derivative is a function of both ξ and of y.

By the second definition,

$$\delta F = F[g(x)+\eta(x)] - F[g(x)]$$

$$= \int_0^\infty K(x,y)(g(x)+\eta(x))\,dx - F[g(x)]$$

$$= \int_0^\infty K(x,y)\eta(x)\,dx.$$

So the functional derivative can be identified as

$$\frac{\delta F}{\delta g} = K(x,y),$$

in agreement with the first definition. The first and second definitions are in general equivalent, as can be shown.

Functional derivatives appeared in the section "The Calculus of Variations" unannounced. There the ONE-DIMENSIONAL EULER-LAGRANGE EQUATION was derived from the condition that the variation of a particular integral vanished. Specifically, the functional

$$S[x(t)] = \int_0^T L(x,\dot{x})\,dt$$

was defined for a given function L, and the one-dimensional Euler-Lagrange equation was found to be satisfied by the $x(t)$ that caused $\delta S[x]$ to vanish. The difference $\delta S[x]$ is just the variation $S[x+\eta] - S[x]$ for a small perturbation $\eta(t)$. The one-dimensional Euler-Lagrange equation can equivalently be found from the functional derivative $\delta S[x]/\delta x$:

$$\begin{aligned}\frac{\delta S}{\delta x(t)} &= \lim_{h\to 0}\frac{1}{h}\left(\int_0^T L(x(\tau)+h\delta(t-\tau), \dot{x}+h\dot{\delta}(t-\tau))\,d\tau - L[x]\right)\\
&= \lim_{h\to 0}\frac{1}{h}\left(\int_0^T d\tau\left(h\delta(t-\tau)\frac{\partial L}{\partial x} + L(x,\dot{x}) + h\dot{\delta}\frac{\partial L}{\partial \dot{x}}\right) - L[x]\right)\\
&= \lim_{h\to 0}\frac{1}{h}\int_0^T d\tau\left(h\delta(t-\tau)\frac{\partial L}{\partial x} + h\dot{\delta}\frac{\partial L}{\partial \dot{x}}\right)\\
&= \lim_{h\to 0}\frac{1}{h}\int_0^T d\tau\left(\frac{\partial L}{\partial x} - \frac{d}{d\tau}\left(\frac{\partial L}{\partial \dot{x}}\right)\right)h\delta(t-\tau)\\
&= \frac{\partial L}{\partial x} - \frac{d}{dt}\left(\frac{\partial L}{\partial \dot{x}}\right)\end{aligned}$$

Thus,

$$\frac{\delta S[x]}{\delta x(t)} = \frac{\partial L}{\partial x} - \frac{d}{dt}\left(\frac{\partial L}{\partial \dot{x}}\right),$$

and the extremum is found as the solution of the functional differential equation (which is equivalent to the one-dimensional Euler-Lagrange equation):

$$\frac{\delta S[x]}{\delta x(t)} = 0$$

Compare this result with the ONE-DIMENSIONAL EULER-LAGRANGE EQUATION on page 29.

Functional power series

We noted above that a functional $g[f(x)]$ can be thought of as the limit of a function of N variables $g(f_1, f_2, \ldots, f_N)$. Such a function of N variables can, of course, be expanded in a power series about some point (here taken to be 0):

$$g(f_1,\ldots,f_N) = g(0,0,\ldots,0) + \sum_j \frac{\partial g}{\partial f_j} f_j + \frac{1}{2}\sum_{j,k}\frac{\partial^2 g}{\partial f_j \partial f_k} f_j f_k + \cdots$$

As $g(f_1, f_2, \ldots, f_N) \to g[f(x)]$, the partial derivative terms attain the form $g'(x)\Delta x$, and the sums here approach integrals. The functional $g[f]$ can thus be represented as a sort of power series:

$$g[f(x)] = K_0 + \int_x dx\, K_1(x) f(x) + \frac{1}{2}\int_x \int_y dx dy\, K_2(x, y) f(x) f(y) + \cdots,$$

where

$$K_0 = g[0],$$

$$K_1(x) = \frac{\delta g}{\delta f(x)},$$

$$K_2(x, y) = \frac{\delta^2 g}{\delta f(x) \delta f(y)},$$

etc. Here the second- and higher-order functional derivatives are defined by analogy to ordinary derivatives. The representation of a continuous functional as a power series of this sort is known as a *Volterra series.*

Just as a power series can sometimes be approximated by just the first two terms, so a Volterra series can be truncated. For a continuous functional $g[f(x) + \eta(x)]$, if $\eta(x)$ is sufficiently small, we can write to first order

$$g[f(x) + \eta(x)] \approx g[f(x)] + \int \frac{\delta g}{\delta f(\xi)} \eta(\xi)\, d\xi,$$

just as $f(x + h) \approx f(x) + hf'(x)$.

The Volterra series can be written in a compact form. We use the notation

$$K_n * f^n = \underbrace{\int \cdots \int}_{n} dx_1 \cdots dx_n\, K_n(x_1, \ldots, x_n) f(x_1) \cdots f(x_n)$$

for this n-fold integral and assume that K_n is symmetric. For example, if $K_2(x, x') = K_2(x', x)$, then

$$g[f(x)] = \sum_k \frac{1}{k!} K_k * f^k,$$

with

$$K_k = \frac{\delta^k g[f]}{\delta f^k}.$$

Use of the definition for the functional derivative shows that, as for ordinary derivatives,

$$\frac{\delta}{\delta f(\xi)} K_n * f^n = nK_n * f^{n-1}.$$

Because this n-fold convolution involves only $n - 1$ functions now, that is,

$$nK_n * f^{n-1} = n \underbrace{\int \cdots \int}_{n-1} K_n(x_1, \ldots, \xi) f(x_1) \cdots f(x_{n-1})\, dx_1 \cdots dx_{n-1},$$

the resulting functional derivative is a function of the point ξ at which the argument was perturbed. For example, if we let $\Theta[f] = K_2 * f^2$ and keep terms only to first order in h, then

$$\begin{aligned}\frac{\delta\Theta[f(x)]}{\delta f(\xi)} &= \lim_{h\to 0} \frac{1}{h}(\Theta[f(x) + h\delta(x + \xi)] - \Theta[f(x)])\\ &= \lim_{h\to 0} \frac{1}{h} \int\int dx dx'\, K(x, x') h(f(x)\delta(x' - \xi) + f(x')\delta(x - \xi))\\ &= 2 \int dx\, K(x, \xi) f(x)\\ &= 2K_2 * f.\end{aligned}$$

Functional integration

A functional integral $\int G[f]\mathscr{D}f$ is defined as the limit of an N-fold integral:

$$\int \mathscr{D}f(x)\, G[f(x)] \equiv \lim_{N\to\infty} \underbrace{\int \cdots \int}_{N} \frac{\prod_1^N df_k}{\theta} g_N(f_1, \ldots, f_N),$$

where $g_N(f_1, \ldots, f_N) \to G[f(x)]$, as described above, and θ is a normalization factor that insures that the integral converges properly. For example, if the functional integral were to represent a probability, θ would be chosen to make the total probability equal to 1. This functional integral is analogous to an ordinary one: the value of the integral is found by taking the value of the functional $F[f]$ for each possible function f in a space of functions, weighting this value by a "length" $\mathscr{D}f$ in function space, and adding them all up. Functional integrals are also called *path integrals* because the integration takes place over all paths $f(x)$ in some particular space.

Consider an example. We wish to evaluate the functional integral

$$K(x,t;\xi,\tau) = \int \mathscr{D}x(t)\exp\left(-\frac{1}{2}\int_{\tau}^{t} \dot{x}^2\,dt'\right),$$

with all paths required to start at $\xi(\tau)$ and end at $x(t)$ and $\dot{x} \equiv dx/dt$. We require that the integral of $K(x,t;\xi,\tau)$ over all x be unity (the normalization condition). We shall evaluate this integral in two ways, one exact and one approximate.

The exact way Divide the time interval $[\tau, t]$ with $N+1$ equally spaced points $t_0 = \tau, t_1, \ldots, t_N = t$ with $\Delta t = t_k - t_{k-1}$. We approximate $x(t')$ with straight line segments between the points $x_k = x(t_k)$. According to the definition of the functional integral K,

$$K(x,t;\xi,\tau) \approx \underbrace{\int\cdots\int}_{N-1} \prod_{1}^{N-1} \frac{dx_k}{\theta}\exp\left(-\frac{1}{2}\sum_{1}^{N}\left(\frac{x_{j+1}-x_j}{\Delta t}\right)^2 \Delta t\right)$$

$$= \underbrace{\int\cdots\int}_{N-1} \prod_{1}^{N-1} \frac{dx_k}{\theta}\exp\left(-\sum_{1}^{N}\frac{\Delta x_k^2}{2\Delta t}\right)$$

$$= \prod_{1}^{N-1}\int_{-\infty}^{\infty}\frac{dx_k}{\theta}\exp\left(\frac{\Delta x_k^2}{2\Delta t}\right),$$

with the approximation becoming exact as $N \to \infty$. Note that the integral contains x_k with k ranging from 1 to $N-1$: x_0 and x_N do not appear because $x_0 = \xi$ is fixed by the initial condition and the integral depends on the variable $x_N = x$. The sum in the exponent (which approximates the integral in the exponent of the functional integral) has the index j, which ranges from 1 to N because the approximate value of the integrand for the kth interval is taken as $(x_k - x_{k-1}/\Delta t)^2$. The normalization factor θ must be determined so that K integrates to 1.

In the section "Random Walk" the probability for a Wiener process moving over an interval Δx in a time step Δt was found to be

$$p(x + \Delta x, t + \Delta t; x, t) = \frac{1}{\sqrt{2\pi\Delta t}} \exp\left(-\frac{(\Delta x)^2}{2\Delta t} \right).$$

And the probability that a diffusion process will move from $\xi(\tau)$ to $x(t)$ in N steps Δt long (because it is a Markov process) is the multiple integral over all of the sequential steps:

$$p(x, t; \xi, \tau) = \frac{1}{\sqrt{2\pi(t - \tau)}} \exp\left(-\frac{(x - \xi)^2}{2(t - \tau)} \right)$$

$$= \underbrace{\int \cdots \int}_{N-1} \prod_{1}^{N-1} dx_k \, p(x_1, t_1; x_0, t_0) \cdots p(x_N, t_N; x_{N-1}, t_{N-1})$$

$$= \underbrace{\int \cdots \int}_{N-1} \prod_{1}^{N-1} \frac{dx_x}{\sqrt{2\pi\Delta t}} \exp\left(-\sum_{1}^{N} \frac{\Delta x^2}{2\Delta t} \right)$$

Here $p(x_{k+1}, t_{k+1}; x_k, t_k)$ is the Gaussian function arising from the random walk, as indicated just above. This is the N-dimensional version of the equation

$$p(x, t; \xi, \tau) = \int p(x, t; x', t') p(x', t'; \xi, \tau) \, dx',$$

which says that the probability of going from $\xi(\tau)$ to $x(t)$ is the sum of all the ways of getting there that pass through the intermediate point $x'(t')$.

This equation for $p(x, t; \xi, \tau)$ expressed as an N-fold integral is the same as the one we wish to evaluate for the functional integral for $K(x, t; \xi, \tau)$ if θ is taken to be $\sqrt{2\pi\Delta t}$, an identification needed to make K integrate to 1. As $N \to \infty$, the preceding equation for $p(x, t; \xi, \tau)$ approaches the function integral for $K(x, t; \xi, \tau)$, so that

$$K(x, t; \xi, \tau) = \frac{\exp(-(x - \xi)^2/2(t - \tau))}{\sqrt{2\pi(t - \tau)}}.$$

The approximate way The function $\int_\tau^t \dot{x}^2 \, dt'$ has a relatively small value for some paths and a much larger value for others. Clearly, because the integral appears in a negative exponential function, the paths that give a small value contribute most to

$$K(x, t; \xi, \tau) = \int \mathscr{D}x(t) \exp\left(-\frac{1}{2}\int_{\tau}^{t} \dot{x}^2\, dt'\right),$$

and the paths with large values contribute little. Specifically, the largest contribution comes from paths near the one for which

$$\frac{\delta}{\delta \dot{x}} \int_{\tau}^{t} \dot{x}^2\, dt' = 0,$$

the condition for a minimum. If we discard terms in h^2, then according to the definition of functional derivatives,

$$\begin{aligned}
\frac{\delta}{\delta \dot{x}} \int_{\tau}^{t} \dot{x}^2\, dt' &= \lim_{h \to 0} \frac{1}{h}\left(\int_{\tau}^{t} (\dot{x} + h\dot{\delta}(t' - t_0))^2\, dt' - \int_{\tau}^{t} \dot{x}^2\, dt'\right) \\
&= \lim_{h \to 0} \frac{1}{h}\left(\int_{\tau}^{t} (\dot{x}^2 + 2h\dot{\delta}(t' - t_0)\dot{x})\, dt' - \int_{\tau}^{t} \dot{x}^2\, dt'\right) \\
&= 2\int_{\tau}^{t} \dot{x}(t')\dot{\delta}(t' - t_0)\, dt' \\
&= -2\int_{\tau}^{t} \ddot{x}(t')\delta(t' - t_0)\, dt' \\
&= -2\ddot{x}.
\end{aligned}$$

(The next to last step used integration by parts.) Thus for the most likely path $x_m(t')$, the one that makes the integral in the exponential function a minimum, $\ddot{x}_m = 0$. This means that

$$x_m(t') = At' + B \quad \text{and} \quad \dot{x}_m = A,$$

where A is a constant. Because $\xi = A\tau + B$ and $x = At + B$ (the starting and ending points of every path), $A = (x - \xi)/(t - \tau)$, so that for the most likely path,

$$\dot{x}_m = \frac{x - \xi}{t - \tau}.$$

Note that $\dot{x}_m$ is constant because the start and end of all paths are fixed. We use this most likely value for $\dot{x}_m$ as an approximation to give

$$
\begin{aligned}
K(x,t;\xi,\tau) &= \int \mathscr{D}x(t)\exp\left(-\frac{1}{2}\int_{\tau}^{t}\dot{x}^2\,dt'\right)\\
&\approx \int \mathscr{D}x(t)\exp\left(-\frac{1}{2}\int_{\tau}^{t}A^2\,dt'\right)\\
&= \int \mathscr{D}x(t)\exp\left(-\frac{(x-\xi)^2}{2(t-\tau)^2}\int_{\tau}^{t}dt'\right)\\
&= \exp\left(-\frac{(x-\xi)^2}{2(t-\tau)}\right)\int \mathscr{D}x\\
&\sim \exp\left(-\frac{(x-\xi)^2}{2(t-\tau)}\right).
\end{aligned}
$$

When normalized to make the integral of K equal to 1, this approximate answer is actually the exact one. This method of approximating functional integrals can be very useful.

Gaussian Random Processes*

Gaussian random processes have important applications and illustrate some uses of the functional techniques described in the preceding section. The purpose of this section is to develop the path-integral representation of these processes.

I begin by reviewing a number of properties of Gaussian random processes. Consider a Gaussian random process $x(t)$ that takes on the values $x_1, x_2, \ldots, x_N$ at the times $t_1, \ldots, t_N$. The probability density $p(x_1, \ldots, x_N)$ for this path is, by definition,

$$
p(x_1, x_2, \ldots, x_N) = \frac{1}{\sqrt{(2\pi)^N |\mathbf{M}|}}\exp\left(-\frac{1}{2}\sum_{i,j} b_{ij}x_i x_j\right),
$$

where $\mathbf{B} = [b_{ij}] = \mathbf{M}^{-1}$ and $\mathbf{M}$ is the covariance matrix. The matrix $\mathbf{B}$ is symmetric; the mean of this process is, for convenience, assumed to be 0. The covariance matrix is $\mathbf{M} = [m_{ij}] = E\{x_i x_j\}$ (I denote expectation with $E\{\ \}$). Because $\mathbf{B}$ is the inverse of $\mathbf{M}$, $\mathbf{MB} = I$, so $\sum_k b_{ik}m_{kj} = \delta_{ij}$. The goal is to show that as $N \to \infty$, the probability of the path $x(t)$ approaches the (nonnormalized) functional

$$P[x(t)] = \exp\left(-\frac{1}{2}\int\int dt dt' \, x(t)x(t')B(t-t')\right)$$

for an even function $B(t)$. The most convenient way to develop this and related equations is to use the *characteristic functional*, that is, the functional Fourier transform of the probability functional $P[x]$. I turn now to a description of the functional Fourier transform.

The functional Fourier transform*

Just as the Fourier transform

$$\mathscr{F}[p(x_1,\ldots,x_N)] \equiv \hat{p}(\xi_1,\ldots,\xi_N)$$

of the function $p(x_1,\ldots,x_N)$ is

$$\mathscr{F}[p(x_1,\ldots,x_N)] = \int_{-\infty}^{\infty} p(x_1,\ldots,x_N)\exp\left(-i\sum_k x_k\xi_k\right)d^N x$$
$$= \hat{p}(\xi_1,\ldots,\xi_N),$$

so we can define the *functional Fourier transform:*

FUNCTIONAL FOURIER TRANSFORM

$$\mathscr{F}\{P[x(t)]\} = \int \mathscr{D}x(t)\, P[x(t)]\exp\left(-i\int dt\, x(t)\xi(t)\right) \equiv \chi[\xi(t)]$$

Here $\chi[\xi(t)]$ is the Fourier transform of $P[x(t)]$, and the subscript k in the N-dimensional Fourier transform has been replaced by the variable t. Note that $\chi[\xi(t)]$ is also, by definition, the expectation $E\{\exp(-i\int dt\, x(t)\xi(t))\}$ when the functional being transformed $P[x(t)]$ is a probability density; $\chi[\xi]$ is the characteristic functional associated with $P[x]$. The functional Fourier transform shares properties with the usual Fourier transform. For example,

$$\mathscr{F}\{P[x-\mu]\} = \exp\left(-i\int \mu(t)\xi(t)\,dt\right)\mathscr{F}\{P[x(t)]\},$$

where μ is a given function. This is like the following property of the ordinary Fourier transform:

$$\mathscr{F}[g(t-a)] = \exp(-ia)\mathscr{F}[g(t)]$$

The shift relationship follows directly from the definition of the functional Fourier transform:

$$\begin{aligned}
\mathscr{F}\{P[x(t)-\mu]\} &= \int \mathscr{D}x\, P[x-\mu] \exp\left(-i\int \xi x\, dt\right) \\
&= \int \mathscr{D}x'\, P[x'] \exp\left(-i\int (x'+\mu)\xi\, dt\right) \\
&= \exp\left(-i\int \xi\mu\, dt\right) \int \mathscr{D}x'\, P[x'] \exp\left(-i\int x'\xi\, dt\right) \\
&= \exp\left(-i\int \xi\mu\, dt\right) \mathscr{F}\{P[x(t)]\},
\end{aligned}$$

where the change of variables $x' = x - \mu$ has been used in the second step. Recall that the function $\mu(t)$ is a constant in function space with respect to the path integral; that is, $\mathscr{D}x' = \mathscr{D}(x-\mu) = \mathscr{D}x$.

To find the characteristic functional that corresponds to a Gaussian process, we need to be able to calculate the functional Fourier transform of

$$P[x(t)] = \exp\left(-\frac{1}{2}\int\int dt dt'\, x(t)x(t')B(t-t')\right).$$

(Note that the probability-density functional has not been normalized; normalization will be implicit throughout.) The best approach is to start with the characteristic function corresponding to an N-dimensional Gaussian process.

The N-dimensional Gaussian probability*

Consider the N-dimensional characteristic function

$$\chi(\xi_1,\ldots,\xi_N) = \exp\left(-\frac{1}{2}\sum_{j,k} m_{jk}\xi_j\xi_k\right),$$

where we require the matrix $\mathbf{M} = [m_{jk}]$ to have positive eigenvalues and to be symmetric ($m_{jk} = m_{kj}$). The first order of business is to find the N-dimensional probability density that corresponds to this characteristic function; it will turn out to be Gaussian.

To find the probability density $p(x_1, \ldots, x_N)$ associated with this characteristic function, we require the inverse Fourier transform:

$$p(x_1, \ldots, x_N) = \frac{1}{(2\pi)^N} \int_{-\infty}^{\infty} \exp\left(-\frac{1}{2} \sum_{j,k} m_{jk} \xi_j \xi_k + i \sum_j \xi_j x_j \right) d^N \xi$$

The calculations are facilitated by defining vectors as $\boldsymbol{\xi} = (\xi_j)$ and $\mathbf{x} = (x_j)$, so that the terms in the exponential function of the preceding integral can be written compactly as

$$-\frac{1}{2} \sum_{j,k} m_{jk} \xi_j \xi_k + i \sum_j \xi_j x_j = -\frac{1}{2} \boldsymbol{\xi} \mathbf{M} \boldsymbol{\xi} + i \boldsymbol{\xi} \mathbf{x}.$$

The $\boldsymbol{\xi}$ and $\mathbf{x}$ can be either row or column vectors, as indicated by the context.

The strategy is to make a change of variables, $\boldsymbol{\xi}$ to $\boldsymbol{\xi}' = \boldsymbol{\xi} - \boldsymbol{\alpha}$, and to choose α so that the argument of the exponential function becomes

$$-\tfrac{1}{2} \boldsymbol{\xi} \mathbf{M} \boldsymbol{\xi} + i \boldsymbol{\xi} \mathbf{x} = -\tfrac{1}{2} \boldsymbol{\xi}' \mathbf{M} \boldsymbol{\xi}' + \text{constants (independent of } \boldsymbol{\xi}').$$

If this form can be achieved, then $\mathbf{M}$ can be diagonalized, and the integral becomes easy, as will be seen. What $\boldsymbol{\alpha}$ should be chosen? We assume that the correct change of variables can be found and substitute $\boldsymbol{\xi}' + \boldsymbol{\alpha}$ for $\boldsymbol{\xi}$ to find what $\boldsymbol{\alpha}$ must be

$$\begin{aligned} -\tfrac{1}{2} \boldsymbol{\xi} \mathbf{M} \boldsymbol{\xi} + i \boldsymbol{\xi} \mathbf{x} &= -\tfrac{1}{2} (\boldsymbol{\xi}' + \boldsymbol{\alpha}) \mathbf{M} (\boldsymbol{\xi}' + \boldsymbol{\alpha}) + i (\boldsymbol{\xi}' + \boldsymbol{\alpha}) \mathbf{x} \\ &= -\tfrac{1}{2} \boldsymbol{\xi}' \mathbf{M} \boldsymbol{\xi}' - \tfrac{1}{2} \boldsymbol{\alpha} \mathbf{M} \boldsymbol{\alpha} + i \boldsymbol{\alpha} \mathbf{x} - \underbrace{(\boldsymbol{\alpha} \mathbf{M} - i \mathbf{x}) \boldsymbol{\xi}'}_{\text{independent of } \boldsymbol{\xi}'}. \end{aligned}$$

For the indicated term to be independent of $\boldsymbol{\xi}'$, what we need is that $(\boldsymbol{\alpha} \mathbf{M} - i \mathbf{x}) = 0$, so

$$\boldsymbol{\alpha} = i \mathbf{M}^{-1} \mathbf{x} = i \mathbf{B} \mathbf{x} = i \mathbf{x} \mathbf{B}$$

for the matrix $\mathbf{B} = \mathbf{M}^{-1}$. With this value for $\boldsymbol{\alpha}$, the expression in the exponential function for the probability p is then

$$\begin{aligned} p(x_1, \ldots, x_N) &= \frac{1}{(2\pi)^N} \int_{-\infty}^{\infty} \exp\left(-\frac{1}{2} \boldsymbol{\xi}' \mathbf{M} \boldsymbol{\xi}' - \frac{1}{2} \mathbf{x} \mathbf{B} \mathbf{x} \right) d^N \xi' \\ &= \frac{\exp(-\frac{1}{2} \mathbf{x} \mathbf{B} \mathbf{x})}{(2\pi)^N} \int_{-\infty}^{\infty} \exp\left(-\frac{1}{2} \boldsymbol{\xi}' \mathbf{M} \boldsymbol{\xi}' \right) d^N \xi'. \end{aligned}$$

For the argument of the exponential function, these equations used the manipulations

$$-\tfrac{1}{2}\boldsymbol{\xi}'\mathbf{M}\boldsymbol{\xi}' - \tfrac{1}{2}(i\mathbf{xB})\mathbf{M}(i\mathbf{xB}) + i(i\mathbf{xB})\mathbf{x} = -\tfrac{1}{2}\boldsymbol{\xi}'\mathbf{M}\boldsymbol{\xi}' + \tfrac{1}{2}\mathbf{xBx} - \mathbf{xBx}$$

$$= -\tfrac{1}{2}\boldsymbol{\xi}'\mathbf{M}\boldsymbol{\xi}' - \tfrac{1}{2}\mathbf{xBx}.$$

The matrix $\mathbf{M}$ can be diagonalized with eigenvalues λ_j, and the resulting change of coordinates does not alter the value of the integral, because the determinant of the required orthogonal matrix is 1. This means that the probability density is

$$p(x_1,\ldots,x_N) = \frac{\exp(-\tfrac{1}{2}\mathbf{xBx})}{(2\pi)^N}\int_{-\infty}^{\infty}\exp\left(-\frac{1}{2}\sum_j \lambda_j\xi_j^2\right)d^N\xi.$$

This N-dimensional integral is the product of N integrals of the form

$$\int_{-\infty}^{\infty}\exp\left(-\frac{1}{2}\lambda_j\xi_j^2\right)d\xi_j = \sqrt{2\pi/\lambda_j}.$$

Thus

$$p(x_1,\ldots,x_N) = \frac{\exp(-\tfrac{1}{2}\mathbf{xBx})}{(2\pi)^N}\prod_1^N\sqrt{2\pi/\lambda_j}.$$

Because $|\mathbf{M}| = \prod_1^N \lambda_j$, the probability density is

$$p(x_1,\ldots,x_N) = \frac{\exp(-\tfrac{1}{2}\mathbf{xBx})}{\sqrt{(2\pi)^N|\mathbf{M}|}}.$$

In summary, the *Gaussian characteristic function* corresponds to the *Gaussian probability density* with the covariance matrix $\mathbf{M} = [m_{jk}] = \mathbf{B}^{-1} = [b_{jk}]^{-1}$:

GAUSSIAN CHARACTERISTIC FUNCTION

$$\chi(\xi_1,\ldots,\xi_N) = \exp\left(-\sum_{j,k} m_{jk}\xi_j\xi_k\right)$$

GAUSSIAN PROBABILITY DENSITY

$$p(x_1,x_2,\ldots,x_N) = \frac{1}{\sqrt{(2\pi)^N|\mathbf{M}|}}\exp\left(-\frac{1}{2}\sum_{i,j} b_{ij}x_ix_j\right)$$

The processes considered above have 0 means. If the mean is μ, then the probability density is

$$p(x_1, x_2, \ldots, x_N) = \frac{1}{\sqrt{(2\pi)^N |\mathbf{M}|}} \exp\left(-\frac{1}{2} \sum_{i,j} b_{ij}(x_i - \mu)(x_j - \mu) \right),$$

and the characteristic function, according to the shift property of the Fourier transform, is

$$\chi(\xi_1, \ldots, \xi_N) = \exp\left(i \sum_j \mu_j \xi_j - \sum_{j,k} m_{jk} \xi_j \xi_k \right).$$

The Gaussian probability functional*

The preceding discussion dealt with the N-dimensional Gaussian. As $N \to \infty$, the sums in the preceding expressions become integrals, and expression for the Gaussian probability functional and characteristic functional result. Specifically, as $\Delta t \to 0$, the covariance function approaches $M(t - t')\Delta t$, so that $\hat{p}(\xi) \to \chi[\xi(t)]$:

$$\chi[\xi(t)] = \exp\left(-\frac{1}{2} \int\int M(t - t') \xi(t) \xi(t') \, dt dt' \right)$$

In the same way the probability density approaches the (nonnormalized) functional:

$$P[x(t)] = \exp\left(-\frac{1}{2} \int\int B(t - t') x(t) x(t') \, dt dt' \right)$$

Further, the relationship $\sum m_{ij} b_{jk} = \delta_{ik}$ approaches

$$\int M(t - \tau) B(\tau - t') \, d\tau = \delta(t - t');$$

that is, the covariance function M is the functional inverse of B. For a nonzero mean $\mu(t)$, these two relationships become the *Gaussian characteristic functional* and the *Gaussian probability density functional*:

GAUSSIAN CHARACTERISTIC FUNCTIONAL

$$\chi[\xi(t)] = \exp\left(i \int \xi(t) \mu(t) \, dt - \frac{1}{2} \int\int M(t - t') \xi(t) \xi(t') \, dt dt' \right)$$

GAUSSIAN PROBABILITY DENSITY FUNCTIONAL

$$P[x(t)] = \exp\left(-\frac{1}{2}\int\int B(t-t')[x(t)-\mu(t)][x(t')-\mu(t')]\,dt\,dt'\right)$$

(Note again that the probability density is not normalized.)

Just as the moments of an N-dimensional process can be calculated from the characteristic function, so the moments of a continuous random process can be found from the characteristic functional. For example, for the Gaussian process discussed above, the mean is given by

$$-\left(i\frac{\delta\chi[\xi(t)]}{\delta\xi(t')}\right)_{\xi=0} = \mu(t'),$$

and the covariance is

$$\left(-\frac{\delta^2\chi[\xi(t)]}{\delta\xi(t)\delta\xi(t')}\right)_{\xi=0} = M(t-t').$$

These relationships arise from the fact that the functional Fourier transform is the expectation

$$\begin{aligned}\chi[\xi(t)] &= E\left\{\exp\left(-i\int dt\, x(t)\xi(t)\right)\right\}\\ &= \int \mathscr{D}x(t)\, P[x(t)]\exp\left(-i\int dt\, x(t)\xi(t)\right).\end{aligned}$$

Thus, for example, the mean of a random process is

$$\begin{aligned}\left(i\frac{\delta\chi[\xi]}{\delta\xi(t')}\right)_{\xi=0} &= \left(i\left[-i\int \mathscr{D}x\, P[x]x\exp\left(-i\int dt\, x(t)\xi(t)\right)\right]\right)_{\xi=0}\\ &= \int \mathscr{D}x\, P[x]x.\end{aligned}$$

The covariance function and higher moments for non-Gaussian processes are found in the same way.

Markov processes*

If the Gaussian random process $x(t)$ described above (assume a mean of 0 for convenience) has the special additional property that its future evolution depends only on its current state, and not on the path it took to reach that state, the process is said to be *Markovian*. A property of a one-dimensional Markov process, known as the Uhlenbeck-Ornstein process, is that the covariance function $C(t)$ is exponential:

$$C(t) = \sigma^2 e^{-|t/\lambda|}$$

Here σ is the standard deviation of the process, and λ is a characteristic time. The characteristic function for this process is thus

$$\chi_M[\xi(t)] = \exp\left(-\frac{1}{2}\int\int dt dt' \, \xi(t)\xi(t')\sigma^2 e^{|t-t'|/\lambda}\right).$$

What is the probability density? Since the function $B(t)$ that will appear in the probability functional is the functional inverse of $C(t)$, that is,

$$C * B = \delta,$$

the most convenient way to find $B(t)$ is to work with Fourier transforms. Let the transform variable be ω. Then

$$\hat{B}(\omega) = \frac{1}{\hat{C}(\omega)},$$

where the transformed functions are indicated with hats. The Fourier transform of $C(t)$ is

$$\hat{C}(\omega) = \int_{-\infty}^{\infty} \sigma^2 e^{-|t|/\lambda - i\omega t}\, dt = \int_{0}^{\infty} \sigma^2 e^{-|t|/\lambda - i\omega t}\, dt + \int_{-\infty}^{0} \sigma^2 e^{-|t|/\lambda - i\omega t}\, dt.$$

Because C is an even function, the last integral on the right (with the change of variable $t = -t'$) becomes

$$\int_{-\infty}^{0} \sigma^2 e^{-|t|/\lambda - i\omega t}\, dt = \int_{0}^{\infty} \sigma^2 e^{-|t|/\lambda + i\omega t}\, dt = \left(\int_{0}^{\infty} \sigma^2 e^{-|t|/\lambda - i\omega t}\, dt\right)^*;$$

that is, the second integral is the complex conjugate of the preceding one. As for the next-to-last integral,

$$\int_{0}^{\infty} \sigma^2 e^{-|t|/\lambda - i\omega t}\, dt = \frac{\sigma^2 \lambda}{1 + i\omega\lambda},$$

so

$$\hat{C}(\omega) = \frac{\sigma^2\lambda}{1 + i\omega\lambda} + \frac{\sigma^2\lambda}{1 - i\omega\lambda} = \frac{2\sigma^2\lambda}{1 + \omega^2\lambda^2}$$

and

$$\hat{B}(\omega) = \frac{1 + \omega^2\lambda^2}{2\sigma^2\lambda}.$$

The problem now is to find the inverse Fourier transform of this last function. A key property of Fourier transforms is that

$$\mathscr{F}\left[\frac{dx(t)}{dt}\right] = i\omega\mathscr{F}[x(t)].$$

Thus $i\omega \Rightarrow d/dt$ and $\omega^2 \Rightarrow -d^2/dt^2$ (where $\Rightarrow$ indicates a correspondence between operators in the Fourier domain and the time domain). The integral in the exponential function of the Gaussian probability functional can be written as

$$\int\int B(t-t')x(t)x(t')\,dtdt' = \int x(t)(B*x)\,dt,$$

and the convolution $B*x$ has the Fourier transform

$$\mathscr{F}[B*x] = \frac{1 + \omega^2\lambda^2}{2\sigma^2\lambda}\hat{x}(\omega).$$

Because $\omega^2 \Rightarrow -d^2/dt^2$, this corresponds to

$$B*x = \frac{x(t) - \lambda^2\ddot{x}}{2\sigma^2\lambda}.$$

The integral is then

$$\int dt\,(B*x)x(t) = \int dt\,x(t)\left(\frac{x(t)-\lambda^2\ddot{x}}{2\sigma^2\lambda}\right) = \int dt\left(\frac{x^2+\lambda^2\dot{x}^2}{2\sigma^2\lambda}\right).$$

The last step used integration by parts to change $-x\ddot{x}$ to $\dot{x}^2$. Finally, then, the (nonnormalized) probability functional for a Gaussian Markov process is

$$P[x(t)] = \exp\left(-\int dt\left(\frac{x^2+\lambda^2\dot{x}^2}{2\sigma^2\lambda}\right)\right).$$

Non-Gaussian processes*

The characteristic functional of a general random process $x(t)$ can be written as

$$\chi[\xi(t)] = \exp(W[\xi(t)])$$

for some functional $W[\xi]$. For Gaussian processes, we know, of course, what this functional is. We can expand $W[\xi]$ in a Volterra series:

$$\chi[\xi(t)] = \exp\left(\int dt\, K_1(t)\xi + \int\int dtdt'\, K_2(t,t')\xi^2 + \int\int\int dtdt'dt''\, K_3(t,t',t'')\xi(t)\xi(t')\xi(t'') + \cdots\right)$$

The first two terms in the expansion give a Gaussian process. As higher-order terms are added, non-Gaussian processes result. By taking functional derivatives of the characteristic function, the moments of this process can be found. The problem, however, is that the inverse functional Fourier transform cannot generally be carried out, so one cannot find the corresponding probability functional.

Cartesian Tensors

A special case of general tensor analysis involves *Cartesian tensors*, those tensors restricted to Cartesian space, which I present in this section. Cartesian tensors are particularly useful in the description of small elastic deformations in electricity and magnetism and in special relativity.

Tensors are natural generalizations of vectors and matrices. They are confusing both because they tend to be messy—components and subscripts abound—and because they do not fit nicely into our concepts of linear operators on vector spaces, although they clearly are somehow related. The most important notion about tensors, their defining characteristic in fact, is that they are invariant under rotations of the space that describe the physical situation. To see what this means, the best starting place is a physical example.

Imagine a plastic body that is slightly deformed; picture a lump of clay that is sagging a little as it sits on a surface. We can identify two arbitrary neighboring points in the material A and B and call the vector from A to B before the deformation $\delta\mathbf{x}$. After the body has changed shape by a slight amount, A will be further displaced from B, so their separation will be altered by an amount $\delta\boldsymbol{\xi}$ (a vector). After the deformation has occurred, the separation between A and B is thus given by $\delta\mathbf{x} + \delta\boldsymbol{\xi}$. How is $\delta\boldsymbol{\xi}$ related to the original vector $\delta\mathbf{x}$ that specifies the relationship between A and B? Suppose that point A has moved under the deformation by amount **r**. Surely the change in separation of A and B will relate to the size of the strain (displacement) **r**. To the first order, this relationship, for the ith components of the vectors, is

$$\delta\xi_i = \sum_j \frac{\partial r_i}{\partial x_j} \delta x_j.$$

We can define a matrix **M** of the partial derivatives with

$$m_{ij} = \frac{\partial r_i}{\partial x_j}.$$

Then the transformation equation can be written as

$$\delta\boldsymbol{\xi} = \mathbf{M}\delta\mathbf{x}.$$

M is an example of a tensor. To see what makes it so, the important thing to note that the deformation of the clay described by **M** has to be just the same, no matter how we orient our coordinate system. If the unitary matrix **C** specifies a certain rotation of the coordinate system used to locate points in the body (under **C**, each vector is transformed $\mathbf{r} \rightarrow \mathbf{r}' = \mathbf{Cr}$), then the deformation would have to be specified in the primed coordinate system as

$$\delta\xi_i' = \sum_j m_{ij}'\delta x_j'.$$

Now transform from the unprimed system to the primed system to get this equation:

$$\delta\boldsymbol{\xi}' = \delta\mathbf{C}\boldsymbol{\xi} = \mathbf{CM}\delta\mathbf{x} = \mathbf{CMC}^{-1}\mathbf{C}\delta\mathbf{x} = \mathbf{CMC}^{-1}\delta\mathbf{x}' = \mathbf{M}'\delta\mathbf{x}'$$

(Remember that $\mathbf{C}^{-1} = \mathbf{C}^*$ for a unitary matrix; see page 12.) So, $\mathbf{M}' = \mathbf{CMC}^{-1}$ and, writing this in components, we get the *definition of a second rank tensor*:

DEFINITION OF
A SECOND-RANK TENSOR

$$m_{ij}' = \sum_{k,l} c_{ik}c_{lj}m_{kl}$$

Any matrix that transforms in this way, that is, for which the components in the rotated coordinate system are related to those in the unrotated coordinate system by this equation, is by definition a tensor. Some matrices have this property—those that represent a valid physical quantity—and they are tensors; the rest of the matrices are not. The key idea here is that any legitimate physical quantity must relate to others in a way that is independent of the orientation of the coordinate system in space.

Some generalizations and extensions of this idea are immediate. The space above was three-dimensional, and the matrix representing the tensor was accordingly 3×3. Clearly, however, space could have more or less than three dimensions, and the preceding discussion would be essentially unchanged. Thus, tensors have a *dimension* that is equal to the number of dimensions of the space in which the physical laws are formulated. Usually this is three, but special relativity, for example, most naturally uses a four-dimensional space, so the tensors in special relativity are represented by 4×4 matrices.

Since the defining characteristic of tensors is the way their components behave when the coordinate system is rotated, one can easily imagine other physical quantities that require not two but some other number of subscripts for each component. Scalars and vectors naturally transform, as do the tensors described above, so they too are included in the family of tensors: all scalars and vectors in a particular space are tensors in that space.

A tensor has not only a dimension but also a *rank* equal to the number of subscripts necessary. The preceding example was a second-rank tensor, a scalar would be a zeroth-rank tensor, and a vector a first-rank tensor. Higher ranks are also sometimes required: in the theory of elasticity, the generalization of Hooke's law uses fourth-rank tensors. A third-rank tensor **T** would transform under a rotation C of the coordinate system according to

$$T'_{ijk} = \sum_{l,m,n} c_{il} c_{jm} c_{kn} T_{lmn}.$$

To compress notation involving tensors, the *Einstein convention* is generally followed: if the same letter appears twice as a subscript, a summation over that subscript is meant. For example,

$$m'_{ij} = a_{ik} a_{lj} m_{kl} \equiv \sum_{k,l} a_{ik} a_{lj} m_{kl},$$

$$x^2 = \mathbf{x} \cdot \mathbf{x} = x_i x_i \equiv \sum_i x_i x_i.$$

A second-rank tensor (with components m_{ij}), like matrices, can be symmetric ($m_{ij} = m_{ji}$), antisymmetric ($m_{ij} = -m_{ji}$), or a mixture. Any second-rank tensor can be separated into symmetric and antisymmetric parts with the relation

$$m_{ij} = \underbrace{\tfrac{1}{2}(m_{ij} + m_{ji})}_{\mathbf{S}} + \underbrace{\tfrac{1}{2}(m_{ij} - m_{ji})}_{\mathbf{A}},$$

where the **S** tensor is symmetric and the **A** tensor antisymmetric. For example, if the components of the antisymmetric tensor **A** are a_{ij}, then

$$a_{ji} = \tfrac{1}{2}(m_{ji} - m_{ij}) = -\tfrac{1}{2}(m_{ij} - m_{ji}) = -a_{ij}.$$

Rotations are represented by unitary matrices, but so are inversions of the coordinate system. Rotations are, of course, the usual physical change in coordinates, but the treatment above holds for any unitary transformation. Some objects that are tensors under rotations do not follow the required transformation equations under inversions. Such "nearly tensors" are termed *pseudotensors.* A pseudotensor transforms under inversions according to the relations given above except that the components are multiplied by a factor det(**C**), where **C** is the transforming unitary matrix. For an inversion of a single axis, det(**C**) $= -1$, so the components of the transformed pseudotensor are all negated by this coordinate change. As long as only rotations are considered, no distinction between tensors and pseudotensors is required.

Summation of tensors is, as would be expected, like that of vectors and matrices: corresponding components are added. Summation is thus defined only for tensors of the same rank and dimension.

Two kinds of tensor multiplication are defined: outer and inner products. For *outer products,* the components of the product tensor are the products of all of the components of the tensors being multiplied. For example, if **A** is a second-rank tensor (with components A_{ij}) and **B** is a first-rank tensor (a vector with components B_i), the outer product of **A** and **B** is a third-rank tensor with components $C_{ijk} = A_{ij}B_k$. This is sometimes written $\mathbf{C} = \mathbf{A} \otimes \mathbf{B}$. A second-rank tensor that is the outer product of two vectors, something that comes up in practice, is designated by the special term *dyad* or *dyadic.*

Inner products, also called *contractions,* are a generalization of the INNER OR DOT PRODUCT familiar from vector spaces. Thus the inner product **C** of a second-rank tensor **A** and a vector **B** is, in terms of its components,

$$C_i = A_{ij}B_j$$

(remember the summation rule). This is, of course, a first-rank tensor (vector). On page 54, a contraction was used to calculate the displacement $\delta\xi$ from the vector that separated two points $\delta\mathbf{x}$ and the tensor **M** that specified the sag of the clay. In general, contractions always have a rank less than that of the highest-rank tensor involved in the contraction.

As stressed earlier, not everything that looks like a tensor, that is, has a bunch of components labeled with subscripts, is a tensor: the test is always how it behaves under unitary transformations of the coordinate space. When attempting to formulate physical laws in a tensorial form (as in Maxwell's equations, for example), one has to demonstrate that the quantities that appear to be tensors actually are. One thing that is a great help in this is the fact that when two of the three quantities involved in an inner product are tensors and the equation relating them is true (that is, based on knowledge of the physical situation), then the third quantity is also a tensor. For example, if we know that **C** and **B** are tensors (say vectors) and **A** is given by the equation **C** = **AB**, then **A** is a tensor of second rank. To see this, apply a unitary matrix **R** to both sides of the equation **C** = **AB** to give

$$\mathbf{RC} = \mathbf{C}' = \mathbf{RAB} = \mathbf{RAR}^{-1}\mathbf{RB} = \mathbf{RAR}^{-1}\mathbf{B}' = \mathbf{A}'\mathbf{B}'.$$

So

$$\mathbf{A}' = \mathbf{RAR}^{-1}$$

or, for the ith component,

$$A'_{ij} = R_{ik}R^*_{lj}A_{kl}.$$

Thus **A** transforms, as it must to be a tensor. We can similarly demonstrate the tensorial character of the third quantity in contractions of tensors of other ranks.

2 Classical Mechanics

The motion of a particle is described by Euler's statement of *Newton's second law*:

NEWTON'S (SECOND) LAW

$$\mathbf{F} = m\mathbf{a}$$

Here $\mathbf{F}$ is the applied force, m is the mass of the particle, and $\mathbf{a} = d\mathbf{v}/dt$ is the particle's acceleration, with $\mathbf{v}$ being the particle's velocity. This equation, together with the principle that bodies act symmetrically on one another—so that the force particle A feels from particle B is equal to the force B feels from A—is the basis for understanding particle dynamics. This chapter summarizes the reformulations of Newton's law that constitute the core theory of classical mechanics.

Newton's law completely describes all of the phenomena of classical mechanics we will consider, but various restatements of this law are used to simplify the solutions of problems or to serve as the basis for theoretical developments in other parts of physics. For example, one statement of classical mechanics is Hamilton's principle, and much of physical theory can be formulated as versions of this principle, as we will see later.

The steps we will follow are (1) to derive the Euler-Lagrange equations for Cartesian coordinates, (2) to develop Hamilton's principle and give the Euler-Lagrange Equations for generalized coordinates, (3) to derive Hamilton's equations, and finally, (4) to present the properties of Poisson brackets. The main goals are to get Hamilton's principle and Hamilton's equations, both of which will play a key role in the treatment of quantum mechanics and field theory.

Euler-Lagrange Equation: First Version

Although Newton's law solves all problems of classical particle dynamics, it is neither the most convenient nor the most general formulation. The problem is that Newton's law is stated in Cartesian coordinates, but certain problems more naturally use other coordinate systems: spherical or cylindrical, for example. More important, many mechanical problems involve forces of constraint, and these are not explicitly known. For example, a roller coaster and a bead sliding on a wire are constrained to move along a particular path, but the forces that dictate that path are complicated and depend on the precise circumstances. Newton's law would require that we know all of the forces acting, all of the time. We need, then, a reformulation of Newton's law that can easily incorporate these forces of constraint (that is, the forces that keep the roller coaster on its track) from their result (e.g., the roller coaster stays on its track) rather than from knowledge in advance of precisely what constraint forces are acting at each point of time and space. What we are after, then, is a generalization of Newton's law that automatically can incorporate forces of constraint. The Euler-Lagrange equation will do this.

To get the first version of the Euler-Lagrange equation, we start with the simplest case of the one-dimensional motion of a single particle with mass m whose position is given by $x(t)$ at time t. The argument given here generalizes at once to a collection of particles and to three dimensions. The particle experiences a conservative force $F(x)$ (like gravity), that is, a force derivable from the scalar potential $V(x)$, where $F = -dV/dx$. Note that in this one-dimensional world, force is a scalar with direction specified by sign.

For a particle of mass m, Newton's law in one dimension is

$$m\ddot{x} - F(x,t) = 0.$$

Note that the force might depend on time as well as position. Three modifications of this equation will lead to the Euler-Lagrange equation.

The first modification is to suppose, as indicated above, that the force can be found from the *potential*, some function of position and possibly of time $V(x,t)$, according to the relation $F(x,t) = -\partial V(x,t)/\partial x$. Substitute this into Newton's law to get

$$m\ddot{x} + \underbrace{\frac{\partial V}{\partial x}}_{-F} = 0.$$

Now notice that use of the particle's momentum $p = m\dot{x}$, an important physical variable, permits Newton's law to be written as

$$\underbrace{\dot{p}}_{m\ddot{x}} + \underbrace{\frac{\partial V}{\partial x}}_{-F} = 0.$$

The trick now is to write the momentum (remember that m is constant) in terms of the particle's kinetic energy:

$$\dot{p} = \frac{d}{dt}(m\dot{x}) = \frac{d}{dt}\underbrace{\frac{\partial}{\partial \dot{x}}\left(\frac{m\dot{x}^2}{2}\right)}_{m\dot{x}} = \frac{d}{dt}\underbrace{\frac{\partial T}{\partial \dot{x}}}_{p}$$

The last step uses the definition of kinetic energy $T = m\dot{x}^2/2$. For the second modification, use this expression to eliminate p from Newton's law, which gives

$$\underbrace{\frac{d}{dt}\frac{\partial T}{\partial \dot{x}}}_{\dot{p}} + \underbrace{\frac{\partial V}{\partial x}}_{-F} = 0.$$

Now notice that the kinetic energy T is a function of $\dot{x}$ but not of x, whereas the potential energy V is a function of x but not of $\dot{x}$ This means we can define a quantity, called the *Lagrangian*, with the property that $\partial L/\partial \dot{x} = \partial T/\partial \dot{x}$ and $\partial L/\partial x = -\partial V/\partial x$:

LAGRANGIAN

$$L(\dot{x}, x) = T(\dot{x}) - V(x)$$

For the final modification, Newton's law can be rewritten as the *Euler-Lagrange equation*, which describes a particle's motion:

EULER-LAGRANGE EQUATION
(CARTESIAN COORDINATES)

$$\underbrace{\frac{d}{dt}\frac{\partial L(x, \dot{x})}{\partial \dot{x}}}_{m\ddot{x}} - \underbrace{\frac{\partial L(x, \dot{x})}{\partial x}}_{F} = 0$$

So far things seem only to have gotten worse. But two simplifications that make it all worthwhile are close at hand.

Hamilton's Principle

Here is the first great advantage of the Euler-Lagrange approach. A comparison of the Euler-Lagrange equation immediately preceding with the ONE-DIMENSIONAL EULER-LAGRANGE EQUATION derived in the section "The Calculus of Variations" (page 27) reveals they are identical. This means that the path $x(t)$ that satisfies the Euler-Lagrange equation is also the one that makes the function S, called *the action*, a minimum:

THE ACTION

$$S[x(t)] = \int_{t_i}^{t_f} L(x, \dot{x})\, dt$$

Thus we have *Hamilton's principle*:

HAMILTON'S PRINCIPLE

$$\delta S = \delta \int_{t_i}^{t_f} L(x, \dot{x})\, dt = 0$$

And $L = T - V$ for the actual path $x(t)$. The criterion for the actual path of a particle $x(t)$ is that the action associated with that path be a minimum. This condition permits us to find a particle's trajectory. Hamilton's principle is a very compact and aesthetic formulation of classical mechanics (you can go backward through the preceding argument from Hamilton's principle to the Euler-Lagrange equation to Newton's law), and it is the first great advantage of the Euler-Lagrange approach. But real importance of the principle arises from the fact that the same *variational principle*, with an appropriate Lagrangian, gives classical electricity and magnetism and that an extension of this principle also produces quantum mechanics. These extensions will be given in later chapters.

Multiple Particles in Three Dimensions

Up to this point we have, for simplicity, considered the example of a single particle moving in one dimension. The generalization to multiple particles in three dimensions is, however, immediate. For N particles with positions $\mathbf{r}_1, \mathbf{r}_2, \ldots, \mathbf{r}_N$ and masses $m_1, m_2, \ldots, m_N$, the kinetic energy of the system is the sum of the individual kinetic energies:

$$T = \sum_{k=1}^{N} \frac{m_k \dot{r}_k^2}{2}$$

And the potential energy V depends, in general, on the positions of all of the particles, because they can interact: $V(\mathbf{r}_1, \mathbf{r}_2, \ldots, \mathbf{r}_N)$. Thus the *Lagrangian for a system* depends on the positions and velocities of all the particles:

LAGRANGIAN (SYSTEM)

$$L(\dot{\mathbf{r}}_1, \mathbf{r}_1, \ldots, \dot{\mathbf{r}}_N, \mathbf{r}_N) = T(\dot{\mathbf{r}}_1, \dot{\mathbf{r}}_2, \ldots, \dot{\mathbf{r}}_N) - V(\mathbf{r}_1, \mathbf{r}_2, \ldots, \mathbf{r}_N)$$

The same arguments as before (page 27) lead to the *3N system Euler-Lagrange equations*, one for each particle k, with identical equations in y and z for the y and z coordinates of each particle:

SYSTEM EULER-LAGRANGE EQUATIONS
(CARTESIAN COODRINATES)

$$\frac{d}{dt}\frac{\partial L(x_k, \dot{x}_k)}{\partial \dot{x}_k} - \frac{\partial L(x_k, \dot{x}_k)}{\partial x_k} = 0$$

Furthermore, the same arguments used for the simpler case lead to the *system Hamilton's principle*:

SYSTEM HAMILTON'S PRINCIPLE

$$\delta S = \delta \int_{t_i}^{t_f} L(\dot{\mathbf{r}}_1, \mathbf{r}_1, \ldots, \dot{\mathbf{r}}_N, \mathbf{r}_N)\, dt = 0$$

This principle constitutes the first great advantage of the Euler-Lagrange approach.

Euler-Lagrange Equation: Second Version

The second great advantage for classical mechanics of the Euler-Lagrange approach comes in restating the problem in *generalized coordinates* rather than in the Cartesian coordinates used above. Here is the reason: Many physical problems come with built in constraints. Two easy examples, mentioned earlier, are the roller coaster and a bead sliding on a wire. Although these situations are complicated in three dimensions, the essential problem is one-dimensional and should be easier. That is, the answer we want for the roller coaster or bead is how far it has progressed along its predefined spatial path. Since we know the path, this information specifies the object's three-space position. The forces of constraint are difficult to use explicitly, which is what Newton's law would require, and the natural statement of the problem is not in terms of forces but rather in equations specifying the constraints that can be used to eliminate some of the variables.

Take a specific example. Suppose that we have a small bead moving without friction along a wire under a conservative force with potential V. The position of the bead at any time is given by the radius vector $\mathbf{r}(t)$, and thus we would have to use the three-dimensional version of Newton's law or of the Euler-Lagrange equations; the three-dimensional version of Hamilton's principle would also identify the actual path of the particle over time. What we really want to know, however, is the distance $q(t)$ that the bead has moved along the wire by time t. From the equations that define the path of the wire through space, we could eliminate $\mathbf{r}$ in favor of q in the equation for the bead's action:

$$S = \int_{t_i}^{t_f} dt\, L(\mathbf{r}, \dot{\mathbf{r}}) = \int_{t_i}^{t_f} dt\, L(q, \dot{q})$$

The equations that specify the constraints might be complicated, so $L(q, \dot{q})$ could be quite messy, but Hamilton's principle still holds, and the actual $q(t)$ would be the one for which $\delta S = 0$. In general, of course, the Lagrangian will depend on more than a single generalized coordinate; for a system of N particles, we would start with $6N$ variables and would be left with a number of generalized coordinates not eliminated by the equations of constraint. We can use the calculus of variations (see page 27) to find the Euler-Lagrange equations that correspond to the path described by generalized coordinates that makes $\delta S = 0$:

EULER-LAGRANGE EQUATION (GENERALIZED COORDINATES)

$$\frac{d}{dt}\left(\frac{\partial L(q_k, \dot{q}_k)}{\partial \dot{q}_k}\right) - \frac{\partial L(q_k, \dot{q}_k)}{\partial q_k} = 0$$

We have one such equation for each pair of generalized coordinates q_k and $\dot{q}_k$ that remain after the equations of constraint have been used to eliminate variables. For the single particle (the bead) moving in three dimensions, we had enough equations to reduce the number of coordinates from 6 to 2 (q and $\dot{q}$), but in other situations the number of generalized coordinates required might be 4 (or even 6). If N particles were involved, then the $6N$ coordinates would be reduced by the various constraints to some number less than or equal to $6N$. Thus, we generally need a system of equations, to solve a problem, but a system simpler than would otherwise be required if the constraints were not present.

The preceding discussion of the Euler-Lagrange equation assumed that the force is conservative, so the term that corresponds to force can be written as $\partial L/\partial q = \partial V/\partial q$. If this is the case, the generalized force is just $G = dV/dq$ (note that the generalized G might not have the units of force, but $G\,dq = dV$ would have the units of work, because V does). If V depends on generalized velocity as well as on position, however, the $\partial L/\partial q = \partial V/\partial q$ are no longer the generalized force, and the Euler-Lagrange equations do not hold. If the generalized force happens to depend on velocity in just the right way—in fact, if

$$G \sim \frac{d}{dt}\left(\frac{\partial V}{\partial \dot{q}}\right) - \frac{\partial V}{\partial q}$$

—then the Euler-Lagrange equations would still hold. Whenever generalized force is of this form, the $d(\partial V/\partial \dot{q})/dt$ will simply add to the term contributed by the kinetic energy $d(\partial L/\partial \dot{q})/dt$, and the form of the Euler-Lagrange equations will be preserved.

Although this may seem to be a particularly restrictive condition on nonconservative forces, the force that a charged particle feels when moving in a magnetic field is precisely of this form. This situation is discussed in chapter 3, where the LORENTZ FORCE ON MOVING CHARGES is shown to have the proper characteristics (p. 99).

Hamilton's Equations

A system of N particles is described, we have learned above, by a system of second-order differential equations numbering $3N$ (or less, depending on how many constraints are present). Since first-order equations are usually preferred to second-order ones, replacing these $3N$ second-order equations with $6N$ first-order equations can be desirable. This is just what the Hamilton-Jacobi theory does. The resulting Hamilton's equations are of fundamental importance for the formulation of quantum mechanics.

We have seen that in Cartesian coordinates the momentum p is given by

$$p = m\dot{x} = \frac{\partial}{\partial \dot{x}}\left(\frac{m\dot{x}^2}{2}\right) = \frac{\partial L(x, \dot{x})}{\partial \dot{x}}.$$

So momentum, for generalized coordinates, is defined as

$$p = \frac{\partial L(q, \dot{q})}{\partial \dot{q}}.$$

This momentum p is known, for obscure historical reasons, as the *canonical conjugate* of the corresponding position q. Because $\dot{p} = d(\partial L/\partial \dot{q})/dt$, the Euler-Lagrange equation

$$\underbrace{\frac{d}{dt}\frac{\partial L}{\partial \dot{q}}}_{\dot{p}} = \frac{\partial L}{\partial q}$$

can thus be represented as

$$\dot{p} = \frac{\partial L(q,\dot{q})}{\partial q}.$$

Here we have a first-order equation, but the problem with this representation is that L depends on $\dot{q}$ rather than on p. A mathematical technique called the *Legendre transformation*, well known in some circles, is precisely suited to get p on the right side of the preceding equation. Here note the possibility that L might be an explicit function of t; this could happen if, for example, the potential energy V varies over time. Associate the function $H(q, p, t)$ with $L(q, \dot{q}, t)$ by the *Hamiltonian* relation:

THE HAMILTONIAN (ONE-DIMENSIONAL)

$$H(q,p,t) = \dot{q}p - L(q,\dot{q},t)$$

For a multidimensional problem, the $\dot{q}p$ term becomes a sum over all (q_i, p_i):

THE HAMILTONIAN (MULTIDIMENSIONAL)

$$H(q_i,p_i,t) = \sum_j \dot{q}_j p_j - L(q_i,\dot{q}_i,t)$$

To find Hamilton's equations, we write

$$dH = \frac{\partial H}{\partial p}dp + \frac{\partial H}{\partial q}dq + \frac{\partial H}{\partial t}dt$$

(because H is a function of q, p, and perhaps t). But from the defining equation for the one-dimensional Hamiltonian, we have

$$dH = \overbrace{\dot{q}\,dp + p\,d\dot{q}}^{d(\dot{q}p)} \underbrace{- \overbrace{\frac{\partial L}{\partial \dot{q}}d\dot{q}}^{p\,d\dot{q}} - \frac{\partial L}{\partial q}dq - \frac{\partial L}{\partial t}dt}_{-dL}.$$

Because $p = \partial L/\partial \dot{q}$, the $pd\dot{q}$ term cancels the $-(\partial L/\partial \dot{q})\,d\dot{q} = -pd\dot{q}$ term, and this last equation reduces to

$$dH = \dot{q}\,dp - \underbrace{\frac{\partial L}{\partial q}dq}_{\dot{p}\,dq} - \frac{\partial L}{\partial t}dt.$$

Further, $\dot{p} = \partial L/\partial q$ (see the Euler-Lagrange equation in the form exhibited on the preceding page), so the $(\partial L/\partial q)\,dq$ term in the equation simplifies to $\dot{p}\,dq$, and we have

$$dH = \dot{q}\,dp - \dot{p}\,dq - \frac{\partial L}{\partial t}dt.$$

Compare this to the original expression for dH. This comparison yields Hamilton's equations:

HAMILTON'S EQUATIONS
(ONE-DIMENSIONAL)

$$\dot{q} = \frac{\partial H}{\partial p}$$

$$\dot{p} = -\frac{\partial H}{\partial q}$$

$$\frac{\partial L}{\partial t} = -\frac{\partial H}{\partial t}$$

For a system of N particles, Hamilton's equations now become $3N$ equations for $\dot{p}_k$ and $3N$ equations for $\dot{q}_k$, or fewer, depending on how many generalized coordinates were required to describe the system. Thus each second-order Euler-Lagrange equation has been replaced by a pair of first-order Hamilton's equations.

The Hamiltonian is, in many cases of interest, just the energy of the system. The conditions under which this is true can be seen from the following.

The total time derivative of the Hamiltonian is

$$\frac{dH}{dt} = \underbrace{\frac{\partial H}{\partial q}\dot{q} + \frac{\partial H}{\partial p}\dot{p}}_{=0} + \frac{\partial H}{\partial t}.$$

But the indicated terms on the right cancel each other (use Hamilton's equations to eliminate $\dot{p}$ and $\dot{q}$) to leave

$$\frac{dH}{dt} = \frac{\partial H}{\partial t}.$$

This means that $dH/dt = 0$ if H does not depend explicitly on time, and thus, in these instances, H is constant as the system evolves in time. H is a *constant of the motion.* What constant?

In Cartesian coordinates, $p\dot{x} = m\dot{x}^2 = 2T$, where T is the kinetic energy. For a conservative system,

$$H = p\dot{x} - L = 2T - L = 2T - (T - V) = T + V.$$

Thus H is the total energy of a conservative system.

Poisson Brackets*

Certain problems require investigating the time evolution not of p and q themselves but of some function of the variables p and q. For example, we examined the time evolution of $H(p, q)$ in the preceding paragraph in order to identify the physical significance of this quantity. A particular expression, known as a *Poisson bracket*, facilitates this process and is of central importance in the development of quantum mechanics.

A *one-dimensional Poisson bracket* is defined, for two functions A and B that depend on canonically conjugate variables p and q, as follows:

POISSON BRACKET (ONE-DIMENSIONAL)

$$\{A, B\} = \frac{\partial A}{\partial q}\frac{\partial B}{\partial p} - \frac{\partial A}{\partial p}\frac{\partial B}{\partial q}$$

For a multidimensional situation, the ps and qs are subscripted, and the terms involving them become sums over all subscripts. That is, we have the *multidimensional Poisson bracket*:

POISSON BRACKET (MULTIDIMENSIONAL)

$$\{A, B\} = \sum_j \left(\frac{\partial A}{\partial q_j}\frac{\partial B}{\partial p_j} - \frac{\partial A}{\partial p_j}\frac{\partial B}{\partial q_j}\right)$$

To see the utility of Poisson brackets, look at the time derivative of a function $A(q, p, t)$:

$$\begin{aligned}\frac{dA}{dt} &= \frac{\partial A}{\partial t} + \frac{\partial A}{\partial q}\dot{q} + \frac{\partial A}{\partial p}\dot{p} \\ &= \frac{\partial A}{\partial t} + \frac{\partial A}{\partial q}\frac{\partial H}{\partial p} - \frac{\partial A}{\partial p}\frac{\partial H}{\partial q} \\ &= \frac{\partial A}{\partial t} + \{A, H\}\end{aligned}$$

Here Hamilton's equations have been used to eliminate $\dot{q}$ and $\dot{p}$ in the second step. Thus the time evolution of a function that depends on canonically conjugate variables is found from a differential equation involving a Poisson bracket and the Hamiltonian:

EVOLUTION EQUATION

$$\frac{dA}{dt} = \{A, H\} + \frac{\partial A}{\partial t}$$

For the special cases of $A = q$ and $A = p$, the Poisson bracket gives a compact representation of HAMILTON'S EQUATIONS when H does not depend explicitly on time:

$$\begin{aligned}\dot{q} &= \{q, H\} \\ \dot{p} &= -\{p, H\}\end{aligned}$$

The Poisson brackets can be applied to the variables p and q themselves. For example,

$$\{q, p\} = \frac{\partial q}{\partial q}\frac{\partial p}{\partial p} - \frac{\partial q}{\partial p}\frac{\partial p}{\partial q} = 1,$$

but

$$\{q, q\} = \frac{\partial q}{\partial q}\frac{\partial q}{\partial p} - \frac{\partial q}{\partial p}\frac{\partial q}{\partial q} = 0,$$

Whenever a pair of variables satisfy the relationship

$$\{r, s\} = 1,$$

the r and s are said to be *canonically conjugate*; p and q, as noted on page 66, are canonically conjugate.

Now consider the situation where we have N variables. For example, from the definition of the Poisson bracket, we have

$$\{q_i, q_j\} = \sum_k \left(\frac{\partial q_i}{\partial q_k}\frac{\partial q_j}{\partial p_k} - \frac{\partial q_i}{\partial p_k}\frac{\partial q_j}{\partial q_k}\right) = 0$$

and

$$\{q_i, p_j\} = \sum_k \left(\frac{\partial q_i}{\partial q_k}\frac{\partial q_j}{\partial p_k} - \frac{\partial q_i}{\partial p_k}\frac{\partial q_j}{\partial q_k}\right) = \frac{\partial q_i}{\partial q_i}\frac{\partial p_j}{\partial p_i} = \delta_{ij}.$$

To get the last equation, we used the KRONOCKER DELTA δ_{ij}.

The following relations follow immediately from the definition of the Poisson bracket, as just illustrated:

$$\{q_i, q_j\} = 0$$

$$\{p_i, p_j\} = 0$$

$$\{q_i, p_j\} = -\{p_j, q_i\}$$

$$\{q_i, p_j\} = \delta_{ij}$$

Again, any pair of variables for which the Poisson bracket is 1 is termed *canonically conjugate*. For the generalized positions and momenta, only q_j and its corresponding generalized momentum p_j would be conjugate, and the others would not. That is, p_i is not the canonical conjugate of q_j. These relationships will be used in the development of quantum mechanics in chapter 4.

3 Electricity and Magnetism

This chapter presents a development of Maxwell's equations for the electromagnetic field.

Maxwell's equations describe two fields, electric and magnetic, and their interactions. These are vector fields. So, according to Helmholtz's theorem (page 5), they can be specified if both their div and curl are known. The goal will be, then, to develop differential equations specifying the div and curl of both electric and magnetic fields. The development takes place in three steps.

The electrostatic field. Coulomb's law is stated as an inverse square relation that specifies the electrostatic field arising from a charge. In principle, the original form of Coulomb's law (the law that describes the total electrostatic field as a linear combination of the contributions from all charges present) solves all electrostatic problems, because the electrostatic field can always be calculated from the charge distribution that generates it. In practice, however, the theory needs to be recast as a pair of differential equations for two reasons. First, the entire charge distribution needed for the application of Coulomb's law is not always available; some problems naturally specify *part* of the charge distribution *and* certain boundary conditions, and a differential formulation is natural for solving these problems. Second, the differential equations are required for later theoretical developments, for example, for Maxwell's equations. Coulomb's law thus must be reformulated as a pair of differential equations for the div and curl of the electrostatic field.

The magnetostatic field. The equations governing the magnetostatic field in a vacuum are, as will be seen later, based on the definition of magnetic-field strength with a small test magnet and two experimental observations. The first of these experimental observations asserts the nonexistence of magnetic monopoles, and the second (Ampère's law) gives the magnetic field associated with a current. These experimental bases for magnetostatics give integral relations that can then be transformed into equations for the div and curl of the magnetostatic field. Differential equations are required here for the same reasons noted above for the electrostatic field.

Time-varying electromagnetic fields. When the electric and magnetic fields are time-varying, rather than stationary, a coupling between the two fields is revealed: the relationship between the electric and magnetic fields comes from Faraday's law, which establishes how a changing magnetic field produces an electric field. When this law is used to extend the div and curl equations for the magnetic and electric fields, four equations result that should specify the electromagnetic field. Maxwell noticed, however, that one of these equations (Ampère's law) is incompatible with the continuity equation expressing the conservation of electric charge. When charge conservation is included in Ampère's law, Maxwell's equations result.

In summary, Maxwell's equations are derived from Coulomb's law, Ampère's law, Faraday's law, charge conservation, and the nonexistence of magnetic monopoles.

In addition to Maxwell's equations, certain extensions are needed for further theoretical developments that involve the electromagnetic field. These include interactions of the electromagnetic field with material media, the wave equations for the electromagnetic field without sources, and also gauge transformations (defined later).

Every area of physics has some feature that makes things difficult. For electromagnetic theory, the bad thing is that most everything comes down to some vector relations that are much less used in other areas of physics and thus are not always familiar. Among the crucial vector relations used repeatedly are the following:

CURL GRAD VANISHES

$$\nabla \times \nabla V = 0$$

DIV CURL VANISHES

$$\nabla \cdot \nabla \times \mathbf{A} = 0$$

STOKES'S THEOREM

$$\oint_C \mathbf{E} \cdot \mathbf{ds} = \int_S \nabla \times \mathbf{E} \cdot d\boldsymbol{\sigma}$$

GAUSS'S THEOREM

$$\int_V \nabla \cdot \mathbf{E}\, d^3x = \int_S \mathbf{E} \cdot d\boldsymbol{\sigma}$$

Vector analysis must be well in hand before (or while) learning or reviewing electromagnetic theory. Refer to the chapter 1 as necessary for a brief presentation of the vector relations required. A recommendation: be able to write by heart the four vector relations above before proceeding with this chapter.

The Electrostatic Field

Charges interact by attracting or repelling one another. Experiments lead to the concept of an electric field $\mathbf{E}$ produced by a constellation of charges: the presence of this field can be detected (defined) by measuring the force $\mathbf{F}$ exerted on a charge q according to the defining equation

$$\mathbf{F} = q\mathbf{E}.$$

Coulomb found that for a pair of charges q and q' separated by a distance r, the force between them is directed along the line joining the charges and has a magnitude specified by *Coulomb's law*:

COULOMB'S LAW

$$F = \frac{qq'}{r^2}$$

Thus a charge q at the origin of the coordinate system produces the field

$$\mathbf{E}(\mathbf{x}) = \frac{q\mathbf{x}}{|\mathbf{x}|^3}.$$

Because fields are found experimentally to add linearly, a collection of charges q_i located at position $\mathbf{x}_i$ produces the field $\mathbf{E}(\mathbf{x})$ at position $\mathbf{x}$:

$$\mathbf{E}(\mathbf{x}) = \sum_i q_i \frac{(\mathbf{x} - \mathbf{x}_i)}{|\mathbf{x} - \mathbf{x}_i|^3}$$

If the charge distribution ρ is continuous, this sum can be written as an integral to give the *generalized Coulomb's law*:

GENERALIZED COULOMB'S LAW

$$\mathbf{E}(\mathbf{x}) = \int \rho(\mathbf{x}') \frac{(\mathbf{x} - \mathbf{x}')}{|\mathbf{x} - \mathbf{x}'|^3} d^3x'$$

If ρ is known, this law will solve all electrostatic problems. In many instances, however, only part of the charge distribution is known, and the rest of the problem is specified by boundary conditions; for example, part of the charge may be known for a particular region, and the electric field might have some specified value on the boundaries of that region. To solve such problems, differential equations are needed. According to Helmholtz, the div and curl of $\mathbf{E}$ must be known, so two equations are required to define $\mathbf{E}$.

Calculations in electrostatics are greatly facilitated by defining an electrostatic potential V from which the field can be derived. This definition will lead at once to the required equations. The key relation is the vector version of $d(1/x)/dx = -1/x^2$:

$$\nabla \frac{1}{|\mathbf{x}-\mathbf{x}'|} = -\frac{(\mathbf{x}-\mathbf{x}')}{|\mathbf{x}-\mathbf{x}'|^3},$$

where the gradient is taken with respect to the variable $\mathbf{x}$. This relation permits the generalized Coulomb's law to be written as

$$\mathbf{E}(\mathbf{x}) = \int \rho(\mathbf{x}') \frac{(\mathbf{x}-\mathbf{x}')}{|\mathbf{x}-\mathbf{x}'|^3} d^3x' = -\nabla \underbrace{\int \frac{\rho(\mathbf{x}')}{|\mathbf{x}-\mathbf{x}'|} d^3x'}_{V(\mathbf{x})}.$$

We define the scalar potential $V(\mathbf{x})$ as

$$V(\mathbf{x}) = \int \frac{\rho(\mathbf{x}')}{|\mathbf{x}-\mathbf{x}'|} d^3x'.$$

Thus the generalized Coulomb's law can be expressed compactly as

$$\mathbf{E}(\mathbf{x}) = -\nabla V(\mathbf{x}).$$

The next job is find the div and curl of $\mathbf{E}$. The curl is gotten immediately, because curl grad vanishes. Since $\mathbf{E}$ can be written as the gradient of a scalar potential V, its curl must vanish:

$$\nabla \times \mathbf{E} = -\nabla \times \nabla V = 0$$

To find the div relation for $\mathbf{E}$, the plan is to go through the divergence theorem (GAUSS'S THEOREM), which is used in considering the integral of $\mathbf{E}$ over a surface S enclosing a charge q: $\int \mathbf{E} \cdot d\boldsymbol{\sigma}$, where $d\boldsymbol{\sigma}$ is a surface element with its vector pointing outward, normal to the enclosing surface. Look at the surface integral over S of the field produced by a single charge q enclosed within S:

$$\int_S \mathbf{E} \cdot d\boldsymbol{\sigma} = \int_S E \cos\theta \, d\boldsymbol{\sigma} = \int_S \frac{q \cos\theta \, d\boldsymbol{\sigma}}{r^2} = q \int d\Omega = 4\pi q$$

Here θ is the angle between the field vector $\mathbf{E}$ at the surface enclosing the charge and the normal to the surface element $d\boldsymbol{\sigma}$, and r is the distance from the charge to the surface element $d\boldsymbol{\sigma}$. Also, recall the definition of a solid angle, $d\Omega = (\cos\theta/r^2)\, d\boldsymbol{\sigma}$ and the fact that $\int d\Omega = 4\pi$. For a continuous charge distribution ρ within the volume enclosed by S, this becomes *Gauss's law*:

GAUSS'S LAW

$$\int_S \mathbf{E}\cdot d\boldsymbol{\sigma} = 4\pi \int \rho(\mathbf{x})\, d^3x,$$

where the right-hand volume integral is over the entire enclosed volume. GAUSS'S THEOREM says

$$\int_S \mathbf{E}\cdot d\boldsymbol{\sigma} = \int \boldsymbol{\nabla}\cdot\mathbf{E}\, d^3x.$$

This means that the volume integrals in the two preceding equations are equal:

$$\int \boldsymbol{\nabla}\cdot\mathbf{E}\, d^3x = 4\pi \int \rho(\mathbf{x})\, d^3x$$

Because the surface is arbitrary, the integrands are equal, and we have the *differential Coulomb's law*:

DIFFERENTIAL COULOMB'S LAW

$$\boldsymbol{\nabla}\cdot\mathbf{E} = 4\pi\rho$$

Thus the electrostatic field can be described by the following two *equations for the electrostatic field*:

EQUATIONS FOR
THE ELECTROSTATIC FIELD

$$\boldsymbol{\nabla}\cdot\mathbf{E} = 4\pi\rho$$

$$\boldsymbol{\nabla}\times\mathbf{E} = 0$$

In terms of the potential, this becomes *Poisson's equation* (remember that $\mathbf{E} = -\boldsymbol{\nabla}V$):

POISSON'S EQUATION

$$\nabla^2 V = -4\pi\rho$$

When the formulation is in terms of the gradient of the voltage, the relation $\nabla \times \mathbf{E} = 0$ is automatically satisfied because curl grad always vanishes. These equations are thus the differential equivalent of Coulomb's law.

The Magnetostatic Field

The existence of a magnetostatic field is inferred by the behavior of permanent magnets. If **m** is the magnetic dipole of a small test magnet and **T** is the torque it experiences, then the magnetic field **B** is defined by

$$\mathbf{T} = \mathbf{m} \times \mathbf{B}.$$

Note that a small magnetic needle whose magnetic dipole is oriented along the needle axis will align with the field **B** because the torque will vanish when **m** and **B** point in the same direction: $\mathbf{m} \times \mathbf{B} = |\mathbf{m}||\mathbf{B}| \sin\theta$, where θ is the angle between the two vectors **m** and **B** and $\sin 0 = 0$. The magnetostatic field is found experimentally to behave linearly.

Magnetostatics is most conveniently based on experimental observations:

Magnetic monopoles do not exist
The observation that leads to the conclusion that there are no magnetic monopoles is that whenever a magnetic is cut in half, both halves possess both north and south poles. This continuous to be true no matter how many times the magnetic is subdivided.

Ampère's law
A steady current in a wire produces a magnetostatic field that can be measured with a small test magnet. The line integral of the magnetostatic field for any contour around a fine current-carrying wire is proportional to the magnitude of the current and is independent of the contour shape. Thus we have *Ampère's law*:

AMPÈRE'S LAW

$$\oint_C \mathbf{B} \cdot d\mathbf{s} = \frac{4\pi}{c} I$$

The factor $4\pi/c$ (c is the speed of light) is not arbitrary but rather is found from experiment. In principle, this law could be verified by using a small test magnetic needle.

The next step is to reformulate these observations as differential equations, as we did for electrostatics. Because **B** is a vector field, it must be specified, as usual, by giving its div and curl. Since magnetic monopoles do not exist,

$$\nabla \cdot \mathbf{B} = 0.$$

The reason div **B** must vanish is that div is a measure of the local density of field sources; the net density must be 0 in any small region because that region, however small, must always contain equal numbers of north and south magnetic poles.

To get the curl equation, use Ampère's law to write the current I flowing through a surface (with surface element $d\boldsymbol{\sigma}$) defined by the contour as

$$\oint \mathbf{B} \cdot d\mathbf{s} = \frac{4\pi}{c} I = \frac{4\pi}{c} \int \mathbf{J} \cdot d\boldsymbol{\sigma}.$$

Here **J** is the current density. STOKES'S THEOREM is that $\oint \mathbf{B} \cdot d\mathbf{s} = \int_S \nabla \times \mathbf{B} \cdot d\boldsymbol{\sigma}$, so the preceding relation can be written

$$\int_S \nabla \times \mathbf{B} \cdot d\boldsymbol{\sigma} = \frac{4\pi}{c} \int_S \mathbf{J} \cdot d\boldsymbol{\sigma}.$$

Since this is true for arbitrary surfaces,

$$\nabla \times \mathbf{B} = \frac{4\pi}{c} \mathbf{J}.$$

For the magnetostatic field, then, we have the following two equations:

EQUATIONS FOR
THE MAGNETOSTATIC FIELD

$$\nabla \times \mathbf{B} = \frac{4\pi}{c} \mathbf{J}$$

$$\nabla \cdot \mathbf{B} = 0$$

The Electromagnetic Field

In the preceding, $\dot{\mathbf{E}}$ and $\dot{\mathbf{B}}$ have been 0, and the two fields have been independent. When the fields are allowed to vary with time, their interdependence becomes apparent. The experimental basis for the coupling of the two fields is *Faraday's law*:

FARADAY'S LAW

$$\oint_C \mathbf{E}\cdot d\mathbf{s} = -\frac{\partial}{\partial t}\frac{1}{c}\int_S \mathbf{B}\cdot d\boldsymbol{\sigma}$$

Here $d\mathbf{s}$ is a length element of the current-carrying wire, and $d\boldsymbol{\sigma}$ is an element of the surface defined by the loop of wire (like a soap film supported by a wire loop). The left side of the equation represents the electromotive force (EMF) impressed on a loop of wire being threaded by the time-varying field **B**, indicated on the right side. The use of STOKES'S THEOREM then gives

$$\oint_C \mathbf{E}\cdot d\mathbf{s} = \int_S \nabla\times\mathbf{E}\cdot d\boldsymbol{\sigma} = -\partial_t\frac{1}{c}\int_S \mathbf{B}\cdot d\boldsymbol{\sigma}.$$

And since the surface shape is arbitrary, this implies that

$$\nabla\times\mathbf{E} = -\frac{1}{c}\partial_t\mathbf{B}.$$

This equation thus couples the electric and magnetic fields for the time-varying case and replaces the $\nabla\times\mathbf{E} = 0$ equation of electrostatics. Thus the four equations that might—but do *not*, it will turn out—specify the electromagnetic field are

$$\nabla\times\mathbf{E} = -\frac{1}{c}\partial_t\mathbf{B}, \qquad \nabla\times\mathbf{B} = \frac{4\pi}{c}\mathbf{J},$$

$$\nabla\cdot\mathbf{E} = 4\pi\rho, \qquad \nabla\cdot\mathbf{B} = 0.$$

Maxwell noticed, however, that these relationships, as they stand, are inadequate because they violate charge conservation as expressed by the continuity equation. To understand Maxwell's insight, we need to derive the continuity equation from charge conservation.

Start with the divergence theorem applied to current **J**:

$$\int_V \nabla\cdot\mathbf{J}\,d^3x = \underbrace{\int_S \mathbf{J}\cdot d\boldsymbol{\sigma}}_{\text{charge loss/sec}}.$$

The right-hand side of this equation is the total current that flows out (*out* is the convention for the surface integral) through the surface of the volume V that we are considering. But by charge conservation, the net outflow of current must be equivalent to the rate at which charges are lost within the volume:

$$\int_V \nabla \cdot \mathbf{J}\, d^3x = \int_S \mathbf{J} \cdot d\boldsymbol{\sigma} = \underbrace{-\frac{\partial}{\partial t} \int_V \rho(\mathbf{x})\, d^3x}_{\text{charge loss/sec}}.$$

Since the volume over which the integrals are taken is arbitrary, we have the *continuity equation*:

CONTINUITY EQUATION

$$\nabla \cdot \mathbf{J} + \frac{\partial \rho}{\partial t} = 0$$

The problem with charge conservation, Maxwell noted, arises with the equation

$$\nabla \times \mathbf{B} = \frac{4\pi}{c} \mathbf{J}$$

(the differential form of AMPÈRE'S LAW). This problem becomes apparent when the divergence of both sides of the preceding equation is taken:

$$\nabla \cdot (\nabla \times \mathbf{B}) = \nabla \cdot \left(\frac{4\pi}{c} \mathbf{J} \right)$$

$$\nabla \cdot \mathbf{J} = \frac{c}{4\pi} \nabla \cdot (\nabla \times \mathbf{B}) = 0$$

The right side of this equation equals zero because div curl vanishes. By the continuity equation, however, div $\mathbf{J}$ should be $-\partial_t \rho$, so to be consistent with charge conservation, the equation should read

$$\frac{c}{4\pi} \nabla \cdot (\nabla \times \mathbf{B}) = \nabla \cdot \mathbf{J} + \partial_t \rho = 0.$$

Since $\rho = \nabla \cdot \mathbf{E}/4\pi$ (assuming the electrostatic equation is correct for time-varying fields), this last equation can be rewritten as a string of divs:

$$\frac{c}{4\pi}\nabla\cdot(\nabla\times\mathbf{B}) = \nabla\cdot\mathbf{J} + \underbrace{\frac{1}{4\pi}\nabla\cdot\partial_t\mathbf{E}}_{\partial_t\rho}$$

This gives a new equation for curl **B**:

$$\nabla\times\mathbf{B} = \frac{4\pi}{c}\mathbf{J} + \frac{1}{c}\partial_t\mathbf{E}$$

The four equations that specify the electromagnetic field in a vacuum now are the following:

MAXWELL'S EQUATIONS

$\nabla\times\mathbf{E} = -\frac{1}{c}\partial_t\mathbf{B}$	Faraday's law
$\nabla\times\mathbf{B} = \frac{4\pi}{c}\mathbf{J} + \frac{1}{c}\partial_t\mathbf{E}$	Ampère's law and conservation of charges
$\nabla\cdot\mathbf{E} = 4\pi\rho$	Coulomb's law
$\nabla\cdot\mathbf{B} = 0$	No magnetic monopoles

The Macroscopic Maxwell's Equations

The equations just developed apply to any situation for which the precise distributions of charges and currents can be specified. In many applications, however, a modified version of the equations, the MACROSCOPIC MAXWELL'S EQUATIONS, are required. In ordinary situations for which the electromagnetic fields are to be described, these fields pass through, and thus interact with, matter like glass, plastic, and water. We turn now the version of Maxwell's equations appropriate for these circumstances.

The electrostatic field in material media

We begin our investigations by considering the electrostatic field. The starting place is an experiment. Imagine that we measure the force around a charge q placed at the center of the coordinate system (the charge is located at $\mathbf{x} = 0$). Then we immerse the apparatus in, say, oil. What we would find is that Coulomb's law still holds, as before, but all of the fields (that is, the forces measured on a test charge) would be reduced by a factor K. The magnitude of the reduction varies, we would find, according to the nature of the material through which the field is passing and can be quite substantial (for example, the reduction would be about eightyfold for water). Clearly, the theory for the electromagnetic field must be able to deal with this situation.

A first guess might be that charges are somehow reduced by contact with the matter in which they are embedded, but experiments reveal that this explanation is too simple. If the material surrounding the charge generating the field is nonuniform, we find that the field at any point is always proportional to the magnitude of the charge q at $\mathbf{x} = 0$ but that the degree of reduction of the field varies from place to place in the material surrounding the charge. That is, the field is modified by the nature of the material in the space between the charge q that generates the field and the test charge used to measure the field even though the material does not touch either charge. What must be happening, then, is that the field itself interacts with the surrounding material and is being modified by it.

According to COULOMB'S LAW, an electric field arises from charges. So a charge distribution must be induced in the material by the electric field we impose (by placing a charge at the center of the coordinate system), and this "reactive" field adds to the original field and thus modifies it. The creation or destruction of charges is forbidden by the CONTINUITY EQUATION (charge conservation), but positive and negative charges that are present in the material can be displaced slightly to produce a net local, induced charge density $\rho_i(\mathbf{E})$ generated by the field. These induced charges themselves produce a field, and this field adds to the field generated by the charge q to result in the total field that we measure by the force on the test charge. The development of the theory now proceeds in two steps: the first is to examine the consequences of the induced charge distribution $\rho_i(\mathbf{E})$, and the second is to develop a theory that relates ρ_i to the applied field $\mathbf{E}$.

First step For a vacuum, the differential form of COULOMB'S LAW is

$$\nabla \cdot \mathbf{E} = 4\pi\rho.$$

This equation is modified by the induced charge distribution $\rho_i(\mathbf{E})$, which adds to the experimentally prescribed charge distribution ρ to give the differential form of COULOMB'S LAW in the presence of material media:

$$\nabla \cdot \mathbf{E} = 4\pi\rho \underset{\substack{\text{material}\\\text{media}}}{\Longrightarrow} \nabla \cdot \mathbf{E} = 4\pi(\rho + \rho_i(\mathbf{E}))$$

Since we cannot know the induced charge distribution—we have direct experimental control only over in "actual" charge distribution ρ—the sensible thing in to reformulate the preceding equation to eliminate ρ_i. We do this in a way that we anticipate will be the best for developing theories that relate the electric field **E** to the induced charge distribution, and this means reformulating ρ_i to give a variable that has the same character as **E**. We start with the definition of *polarization*:

POLARIZATION

$$\nabla \cdot \mathbf{P} = -\rho_i$$

Here **P** is the polarization vector, whose meaning we will examine a little later. Now, though, we can see that **P** is the negative of the electric field (divided by the constant factor 4π) that is produced by the induced charges, and thus is a variable of the same sort (that is, a vector with the same units) as the field **E** that generates it. We add the polarization to the differential form of COULOMB'S LAW and note that the induced charge distribution cancels:

$$\nabla \cdot \underbrace{(\mathbf{E} + 4\pi\mathbf{P})}_{\mathbf{D}} = 4\pi\rho$$

In effect, we have moved the unknown induced charge density from the right to the left side of the equation by the way we construct **P**. The advantage is that **P** has the same character as **E** and will, we will see, be easier to deal with. We now define **D**, the *electric displacement*:

ELECTRIC DISPLACEMENT

$$\mathbf{D} = \mathbf{E} + 4\pi\mathbf{P}$$

In terms of this defined field, the DIFFERENTIAL COULOMB'S LAW becomes the *macroscopic differential Coulomb's law*:

MACROSCOPIC DIFFERENTIAL
COULOMB'S LAW

$$\nabla \cdot \mathbf{D} = 4\pi\rho$$

COULOMB'S LAW is written here in terms of the displacement vector **D**, but the "real" electric field is **E**, not **D**, because **E** is what would be measured by the force on a small test charge. The macroscopic differential Coulomb's law is of the same form as the microscopic law but is an incomplete treatment of electric fields in material media because we cannot find the actual field **E** without a theory that relates **D** (or, equivalently, **P** or ρ_i) to **E**. The second step is to develop this theory.

Second step Roughly speaking, matter can be divided into two classes: *conductors* and *insulators* (also called *dielectrics*). Couductors, like wires, have a practically unlimited quantity of electrons that are free to be transported any distance within the material. In contrast, insulators have virtually no such free electrons; all of their electrons are tightly bound to the atoms making up the material. Although the charges in materials of this sort cannot be transported through the material, they can be displaced slightly from their resting positions, and these displacements produce the charge distribution $\rho_i(\mathbf{E})$ and consequently the polarization $\mathbf{P}(\mathbf{E})$.

For most situations, the simplest possible theory, to be described now, works perfectly. Any field we impose on a dielectric would be small compared to the giant internal fields produced by the charged particles that make up atoms. Thus we can view **E** as a small pertubation and reasonably expand $\mathbf{P}(\mathbf{E})$ in a power series to first order: $\mathbf{P}(\mathbf{E}) = \chi\mathbf{E}$. The coefficient of the linear term, χ is called the *electric susceptibility*, and the zero-order constant term does not appear, because no resting polarization is present in ordinary materials.

The definition of the ELECTRIC DISPLACEMENT vector,

$$\mathbf{D} = \mathbf{E} + 4\pi\mathbf{P},$$

together with the expression for **P** that arises in the power series expansion, $\mathbf{P}(\mathbf{E}) = \chi\mathbf{E}$, provide the required relation between **D** and **E**:

$$\mathbf{D} = \mathbf{E} + 4\pi\mathbf{P} = \mathbf{E} + 4\pi\chi\mathbf{E} = \underbrace{(1 + 4\pi\chi)}_{K}\mathbf{E} \equiv K\mathbf{E}$$

Here K is the *dielectric constant* that appeared at the outset of this discussion.

What does the **P** vector mean physically? Because $\nabla \cdot \mathbf{P} = -\rho_i$ and the units on the right side of the equation are charge/volume, the units of **P** are (charge/length)/volume (since $\nabla\cdot$ is a derivative with respect to length).

Charge/length specifies the units of dipole moment, so polarization **P** is the dipole density induced by the field. This makes sense in terms of our picture of charge separation occurring locally as a result of interactions between **E** and atoms in the dielectric.

The magnetostatic field in material media

The same arguments just given for the electrostatic field can be used for the magnetostatic field. Magnetostatic fields are generated by charge movements, but the fields themselves can, in some types of material, generate current flow that is not directly under experimental control. By experiment, magnetic monopoles still do not exist for the macroscopic case, so $\nabla \cdot \mathbf{B} = 0$, as before. The existence of a **B** field can, however, produce tiny current loops in some materials, so the differential form of AMPÈRE'S LAW becomes

$$\nabla \times \mathbf{B} = \frac{4\pi}{c}\mathbf{J} \xrightarrow[\text{media}]{\text{material}} \nabla \times \mathbf{B} = \frac{4\pi}{c}(\mathbf{J} + \mathbf{J}_i),$$

where $\mathbf{J}_i(\mathbf{B})$ is the reactive current produced by field **B**. As before, define a magnetic-field-like vector **M** by the relation for the *magnetization*:

MAGNETIZATION

$$\nabla \times \mathbf{M} \equiv \frac{1}{c}\mathbf{J}_i$$

Use this equation to eliminate $\mathbf{J}_i$ in the differential form of AMPÈRE'S LAW to give

$$\underbrace{\nabla \times \mathbf{B} - 4\pi(\nabla \times \mathbf{M})}_{\nabla \times \mathbf{H}} = \frac{4\pi}{c}\mathbf{J},$$

a version of AMPÈRE'S LAW that has included the effects of the induced currents in the **M** vector and has kept the experimentally controlled current **J** on the right of the equation. Now we define the **H** *field*:

H FIELD

$$\mathbf{H} \equiv \mathbf{B} - 4\pi\mathbf{M}$$

This definition can be used to simplify the preceding equation to give the *macroscopic differential Ampère's law*:

MACROSCOPIC DIFFERENTIAL
AMPÈRE'S LAW

$$\nabla \times \mathbf{H} = \frac{4\pi}{c}\mathbf{J}$$

This is the magnetostatic analog of the MACROSCOPIC DIFFERENTIAL COULOMB'S LAW, in which **D** is a defined field vector like the **H** here. The "real" field, however, is **B**, because this is what you would measure with a small test magnet.

Because **B** is what must be found, the theory is incomplete until we specify the relation between **B** and **H**(**B**). As with the electrostatic case, the simplest possible theory works perfectly in practice. The same power-series argument used above for finding the relation between **D** and **E** gives, for the magnetic field,

$$\mathbf{B} = \mu\mathbf{H},$$

where μ is called the *magnetic permeability* of the medium being magnetized. This relation—or a more complicated nonlinear one for less simple media—completes the theory. Although the linear theory works well for most of the commonly encountered materials for electrostatics, iron and similar common substances require an additional theory for the magnetostatic field.

The macroscopic Maxwell's equations

This development follows precisely the earlier one that led to Maxwell's equations for the microscopic case (page 78). The equations for the macroscopic fields, including FARADAY'S LAW, are as follows:

$$\nabla \times \mathbf{E} = -\frac{1}{c}\partial_t \mathbf{B} \qquad \nabla \times \mathbf{H} = \frac{4\pi}{c}\mathbf{J}$$

$$\nabla \cdot \mathbf{D} = 4\pi\rho \qquad \nabla \cdot \mathbf{B} = 0$$

As with the vacuum case, the divergence of the MACROSCOPIC DIFFERENTIAL AMPÈRE'S LAW, to be consistent with the CONTINUITY EQUATION, should read

$$\underbrace{\frac{c}{4\pi}\nabla\cdot(\nabla\times\mathbf{H})}_{\text{div curl vanishes}} = \nabla\cdot\mathbf{J} + \partial_t\rho = 0.$$

Now use the MACROSCOPIC DIFFERENTIAL COULOMB'S LAW,

$$\rho = \frac{1}{4\pi}\nabla\cdot\mathbf{D},$$

to eliminate ρ from the preceding equation τ, which will give

$$\frac{c}{4\pi}\nabla\cdot(\nabla\times\mathbf{H}) = \nabla\cdot\mathbf{J} + \frac{1}{4\pi}\nabla\cdot\partial_t\mathbf{D}.$$

This gives the appropriate macroscopic replacement for the differential form of AMPÈRE'S LAW:

$$\nabla\times\mathbf{H} = \frac{4\pi}{c}\mathbf{J} + \frac{1}{c}\partial_t\mathbf{D}$$

We then have the *macroscopic Maxwell's equations*:

MACROSCOPIC MAXWELL'S EQUATIONS

$\nabla\times\mathbf{E} = -\frac{1}{c}\partial_t\mathbf{B}$	Faraday's law
$\nabla\times\mathbf{H} = \frac{4\pi}{c}\mathbf{J} + \frac{1}{c}\partial_t\mathbf{D}$	Ampère's law and conservation of charges
$\nabla\cdot\mathbf{D} = 4\pi\rho$	Coulomb's law
$\nabla\cdot\mathbf{B} = 0$	No magnetic monopoles

To complete the theory, we must have the *accessory relations*:

ACCESSORY RELATIONS

$$\mathbf{D} = \mathbf{D}(\mathbf{E},\mathbf{B})$$

$$\mathbf{H} = \mathbf{H}(\mathbf{E},\mathbf{B})$$

For "usual" cases, these are given by the simplest linear theory:

$$\mathbf{D} = K\mathbf{E}$$

$$\mathbf{H} = \mu\mathbf{B}$$

Behavior of the fields across material boundaries*

If the world were uniformly filled with a simple homogeneous material, the theory above would be unnecessary. All we would have to do is to replace **D** by $K\mathbf{E}$ in the MACROSCOPIC DIFFERENTIAL COULOMB'S LAW, and **H** by $\mu\mathbf{B}$ in the MACROSCOPIC DIFFERENTIAL AMPÈRE'S LAW, and we would be done. Because the world contains objects made of different materials, fields pass through material discontinuities, and this requires that we maintain the distinctions between **D** and **E** and between **H** and **B**.

To discover the behavior of **D** and **E** at the boundaries between dielectrics with different characteristics (i.e., different dielectric constants), we need to go back from the differential relations above to the integral formulations of the laws. Start with the MACROSCOPIC DIFFERENTIAL COULOMB'S LAW: $\nabla \cdot \mathbf{D} = 4\pi\rho$. Integrate both sides over an arbitrary volume to give

$$\int d^3x\, \nabla \cdot \mathbf{D} = \int d^3x\, 4\pi\rho,$$

and apply GAUSS'S THEOREM to the left side to convert the volume integral to a surface integral:

$$\int_S \mathbf{D} \cdot \mathbf{n}\, d\sigma = \int d^3x\, 4\pi\rho$$

Here **n** is the unit normal vector that points out of the cylinder, and $d\sigma$ is an infinitesimal bit of surface area. This relation is, of course, just GAUSS'S LAW for the displacement vector. Now imagine a very short, cylindrical volume with ends of surface area a that sits astride the interface between the two dielectrics; the cylinder is oriented so that its ends are parallel to the interface and it has one end in each dielectric. Such a small volume is called a "Gaussian pillbox."

If we make the sides of the pillbox small enough, the equation above approaches

$$\underbrace{(\mathbf{D}_1 - \mathbf{D}_2) \cdot \mathbf{n}a}_{\int_S \mathbf{D} \cdot \mathbf{n}\, d\sigma} = \underbrace{4\pi\eta a}_{\int d^3x\, 4\pi\rho}.$$

And if we cancel surface area a,

$$(D_{\mathrm{n}_1} = D_{\mathrm{n}_2}) = 4\pi\eta,$$

where D_{n_1} and D_{n_2} are the normal components of the displacement vectors on each side of the dielectric discontinuity and η is the surface charge density. This relation specifies how to deal with the **D** field at dielectric boundaries. How about the **E** field?

To discover the behavior of the **E** field at a dielectric boundary, start with the differential form of FARADAY'S LAW:

$$\nabla \times \mathbf{E} = -\frac{1}{c}\frac{\partial \mathbf{B}}{\partial t},$$

which (if you go backwards through STOKES'S THEOREM) gives

$$\oint \mathbf{E} \cdot d\mathbf{s} = -\frac{1}{c}\int_S \frac{\partial \mathbf{B}}{\partial t} \cdot \mathbf{n}\, d\sigma.$$

Here **n** is the outward directed normal vector for the surface element $d\sigma$. Imagnine now a closed loop that runs parallel to the surface of the first material just next to the boundary, then through the boundary, then parallel to the boundary through the second material, and finally closes by passing through the boundary again. Since the edges that cross the boundary can be made arbitrarily short, the left side of the preceding equation gives

$$\oint \mathbf{E} \cdot d\mathbf{s} = (E_{t_1} - E_{t_2})\,\Delta s$$

for the tangential components of the electric field at a material discontinuity. The right integral vanishes, however, because the area of the loop becomes 0 as the sides become sufficiently short:

$$\int_S \frac{\partial \mathbf{B}}{\partial t} \cdot \mathbf{n}\, d\sigma \to 0$$

The final relation for the tangential components of the **E** field, then, is

$$E_{t_1} - E_{t_2} = 0$$

at the boundary between two dielectrics.

Similar considerations are used for the magnetic field. GAUSS'S THEOREM applied to $\nabla \cdot \mathbf{B} = 0$ gives the integral relation

$$\int_S \mathbf{B} \cdot \mathbf{n}\, d\sigma = 0.$$

Carry this integral out for an arbitrarily shallow Gaussian pillbox (and cancel the surface area) to get

$$B_{n_1} - B_{n_2} = 0$$

for the normal components of the magnetic field.

To find the tangential components of the **H** field, start with AMPÈRE'S LAW (modified to be consistent with charge conservation),

$$\nabla \times \mathbf{H} = \frac{4\pi}{c}\mathbf{J} + \frac{1}{c}\partial_t \mathbf{D},$$

and integrate over a surface to get

$$\int_S \nabla \times \mathbf{H} \cdot d\boldsymbol{\sigma} = \int_S \left(\frac{4\pi}{c}\mathbf{J} + \frac{1}{c}\partial_t \mathbf{D}\right) \cdot d\boldsymbol{\sigma}.$$

Apply STOKES'S THEOREM to end up with the integral form of the modified AMPÈRE'S LAW:

$$\oint \mathbf{H} \cdot d\mathbf{s} = \int_S \left(\frac{4\pi}{c}\mathbf{J} + \frac{1}{c}\partial_t \mathbf{D}\right) \cdot d\boldsymbol{\sigma}$$

Now take this integral around a loop like that used above to find the tangential components of the electric field. The left side of the preceding equation gives, in the limit of a very tight loop, an expression for the tangential components of the **H** field:

$$\oint \mathbf{H} \cdot d\mathbf{s} \rightarrow (H_{t_1} - H_{t_2})\Delta s$$

The integral on the right, which involves $\partial_t \mathbf{D}$, vanishes as the surface over which the integral is carried out becomes 0. But as the sides of the loop become shorter,

$$\frac{4\pi}{c}\int_S \mathbf{J} \cdot d\boldsymbol{\sigma} \rightarrow \frac{4\pi}{c} J_s \Delta s.$$

Here J_s is the surface current density flowing just along the boundary itself; this is the current analog of the surface charge density η that appeared above in the equation for the normal components of **D** as the boundary is crossed.

In summary, the behavior of the quantities **D**, **E**, **H**, and **B** as they cross a discontinuity between material of different characteristics is described by the four equations that specify field behavior at boundaries:

FIELD BEHAVIOR AT BOUNDARIES

$$\Delta D_{\mathrm{n}} = 4\pi\eta$$

$$\Delta E_{\mathrm{t}} = 0$$

$$\Delta H_{\mathrm{t}} = \frac{4\pi}{c} J_s$$

$$\Delta B_{\mathrm{n}} = 0$$

The existence of these relations underlies the requirement that the macroscopic Maxwell's equations contain all four fieldlike variables.

Gauge Transformations*

The description of the electrostatic field was simplified by the introduction of the potential V (see page 74). By its construction, V automatically satisfied the equation curl $\mathbf{E} = 0$ (because curl grad vanishes), so the use of the potential reduced the pair of first-order equations for the electrostatic field to a single second-order equation (Poisson's equation). A similar simplification can be introduced into magnetostatics by using a potential, but in this instance the potential must be a vector (not a scalar). Further, MAXWELL'S EQUATIONS can also be simplified by a similar stratagem. We now examine the simplifications of magnetostatics and electrodynamics that result from a formulation in terms of potentials.

Because div **B** vanishes, **B** can be represented as curl of a vector potential **A** defined by the differential equation

$$\mathbf{B} = \nabla \times \mathbf{A}.$$

Since div curl **A** always vanishes, the relationship div $\mathbf{B} = 0$ is automatically satisfied by this construction, just as curl $\mathbf{E} = 0$ was satisfied by the scalar potential V. The existence of such an **A** can be proved by exhibiting one; this was done in the proof of Helmholtz's theorem. Recall the vector relationship $\nabla \times \nabla \times = \nabla(\nabla \cdot) - \nabla^2$. Thus the curl **B** equation is

$$\nabla \times \mathbf{B} = \nabla \times \nabla \times \mathbf{A} = \nabla(\nabla \cdot \mathbf{A}) - \nabla^2 \mathbf{A}.$$

A itself is, of course, a vector field and thus may be divided into two components, one of which has a vanishing div and the other a vanishing curl. Certainly curl **A** cannot vanish, but we are free to require that div $\mathbf{A} = 0$. For such an **A**, $\nabla \times \mathbf{B} = -\nabla^2 \mathbf{A}$, and the div and curl equations of magnetostatics are expressed as the single-vector equation

$$\nabla^2 \mathbf{A} = -\frac{4\pi}{c}\mathbf{J}.$$

Note that this is really three POISSON EQUATIONS, one for each component. Thus for the x component,

$$\nabla^2 A_x = -\frac{4\pi}{c} J_x.$$

Using the vector potential in this way, one reduces magnetostatic problems to the same form as electrostatics.

Both electrostatics and magnetostatics are thus simplified by introducing the potentials V and **A**. A simplification of the same sort can result for the time-varying case if one defines appropriate generalized vector and scalar potentials. Various potentials will work, and depending on the exact nature of the problem, one or another will give simpler results. Moving between members of the class of acceptable potentials is known, for historical reasons, as carrying out *gauge transformations.* These gauge transformations are useful for getting simpler equations, as we will see below, but are even more important now for applications in other areas of physics.

The generalized vector and scalar potentials are introduced in a way analogous to the one used earlier. Three steps are necessary before the utility of this process is apparent. The first step is to define the generalized potentials, the second, to develop the notion of gauge transformation, and the third, to see how the potentials can be used to simplify MAXWELL'S EQUATIONS.

First step Because $\nabla \cdot \mathbf{B} = 0$ still holds for the time-varying fields described by Maxwell's equations (no magnetic monopoles), the generalized vector potential can be defined, as before, by the differential equation

$$\mathbf{B} = \nabla \times \mathbf{A}$$

(remember that div curl vanishes). For the electrostatic case,

$$\nabla \times \mathbf{E} = 0,$$

so this equation is automatically satisfied if V is defined by the equation

$$\mathbf{E} = -\nabla V,$$

because $\nabla \times \nabla V$ vanishes for any scalar V. For the time-varying case, however, the curl $\mathbf{E}$ equation is more complicated:

$$\nabla \times \mathbf{E} + \frac{1}{c}\partial_t \mathbf{B} = 0$$

But this equation can be written

$$\nabla \times \mathbf{E} + \nabla \times \frac{1}{c}\partial_t \mathbf{A} = 0$$

(use $\mathbf{B} = \nabla \times \mathbf{A}$). So since we already know $\mathbf{A}$ and since curl grad V always vanishes, V can be defined by the differential equation

$$\nabla V = -\left(\mathbf{E} + \frac{1}{c}\partial_t \mathbf{A}\right),$$

because

$$\begin{aligned} \nabla \times \nabla V &= 0 \\ &= -\left(\nabla \times \mathbf{E} + \frac{1}{c}\partial_t \nabla \times \mathbf{A}\right) \\ &= -\left(\nabla \times \mathbf{E} + \frac{1}{c}\partial_t \mathbf{B}\right). \end{aligned}$$

The two differential equations $\nabla V = -(\mathbf{E} + \partial_t \mathbf{A}/c)$ and $\mathbf{B} = \nabla \times \mathbf{A}$, with two unknown functions V and $\mathbf{A}$, can certainly be used to define the generalized potentials so that Maxwell's equations are reduced to the pair for curl $\mathbf{B}$ and div $\mathbf{E}$. Before we can see how such a definition might be useful, we must consider the gauge transformations that define a class of satisfactory potentials.

The gradient of an arbitrary scalar function can always be added to $\mathbf{A}$ without changing the magnetic field (recall yet again that curl grad always vanishes):

$$\mathbf{A} \rightarrow \mathbf{A}' = \mathbf{A} + \nabla M,$$

for an arbitrary function M. This substitution will have no effect on the magnetic field, because $\mathbf{B} = \nabla \times \mathbf{A} = \nabla \times \mathbf{A}' = \mathbf{B}'$, but the added term will potentially change the definition of V, since the definition of V depends on $\mathbf{A}$. In particular, the new $\mathbf{A}'$ will give rise to a $\nabla\partial_t M/c$ term in the definition of V (specifically, $\mathbf{E} = -\nabla \mathbf{V} - \partial_t \mathbf{A}/c$), and thus it will alter $\mathbf{E}$. This can be fixed by simultaneously transforming V and $\mathbf{A}$, so that

$$V \to V' = V - \frac{1}{c}\partial_t M.$$

Now

$$\mathbf{E}' = -\nabla V' - \frac{1}{c}\partial_t \mathbf{A}' = \underbrace{-\nabla V + \frac{1}{c}\nabla\partial_t M}_{V'} \underbrace{- \frac{1}{c}\partial_t \mathbf{A} - \frac{1}{c}\nabla\partial_t M}_{-\partial_t \mathbf{A}'/c} = \mathbf{E}.$$

Second step The family of potentials defined in this way for arbitrary M all lead to the same fields $\mathbf{B}$ and $\mathbf{E}$, and thus are equivalent. It will turn out the freedom to choose M can simplify the equations that describe the electromagnetic field. Thus we have the *definition of potentials* and the *gauge transformations*:

DEFINITION OF POTENTIALS

$$\mathbf{B} = \nabla \times \mathbf{A}$$

$$\mathbf{E} = -\left(\nabla V + \frac{1}{c}\partial_t \mathbf{A}\right)$$

GAUGE TRANSFORMATIONS

$$\mathbf{A} \to \mathbf{A} + \nabla M$$

$$V \to V - \frac{1}{c}\partial_t M$$

Third step With the definitions for $\mathbf{A}$ and V given above, the following two MAXWELL'S EQUATIONS are automatically satisfied:

$$\nabla \cdot \mathbf{B} = \nabla \cdot \nabla \times \mathbf{A} = 0 \quad \text{(no magnetic monopoles)}$$

$$\nabla \times \mathbf{E} = -\nabla \times \left(\nabla V + \frac{1}{c}\partial_t \mathbf{A}\right)$$

$$= -\frac{1}{c}\partial_t \nabla \times \mathbf{A}$$

$$= -\frac{1}{c}\partial_t \mathbf{B} \quad \text{(Faraday's law)}$$

Now look at the remaining two of MAXWELL'S EQUATIONS in terms of potentials:

$$\nabla \times \mathbf{B} = \frac{4\pi}{c}\mathbf{J} + \frac{1}{c}\partial_t \mathbf{E} \quad \text{(Ampère's law and conservation of charges)}$$

$$\nabla \cdot \mathbf{E} = 4\pi\rho \quad \text{(Coulomb's law)}$$

By the definitions of $\mathbf{A}$ and V, we have that

$$\nabla \times \mathbf{B} = \frac{4\pi}{c}\mathbf{J} - \overbrace{\frac{1}{c}\partial_t\left(\nabla V + \frac{1}{c}\partial_t \mathbf{A}\right)}^{\partial_t \mathbf{E}/c}$$

$$= \frac{4\pi}{c}\mathbf{J} - \frac{1}{c^2}\partial_t^2 \nabla \times \mathbf{A} - \frac{1}{c}\nabla \partial_t V$$

(remember again that curl grad vanishes). But also (by the vector relationship $\nabla \times \nabla \times \mathbf{A} = \nabla\nabla \cdot \mathbf{A} - \nabla^2 \mathbf{A}$, given on page 90),

$$\nabla \times \mathbf{B} = \nabla \times \nabla \times \mathbf{A}$$

$$= \nabla\nabla \cdot \mathbf{A} - \nabla^2 \mathbf{A}$$

$$= \frac{4\pi}{c}\mathbf{J} + \frac{1}{c}\partial_t \mathbf{E}$$

$$= \frac{4\pi}{c}\mathbf{J} - \frac{1}{c}\partial_t V - \frac{1}{c^2}\partial_t^2 \mathbf{A}.$$

Now eliminate $\nabla \times \mathbf{B}$ and arrange the $\mathbf{J}$ term to appear on the right. This gives

$$\nabla \times \mathbf{B} = -\nabla^2 \mathbf{A} + \frac{1}{c^2}\partial_t^2 \mathbf{A} + \nabla\underbrace{\left(\nabla \cdot \mathbf{A} + \frac{1}{c}\partial_t V\right)}_{\text{Lorentz gauge}} = \frac{4\pi}{c}\mathbf{J}.$$

Also, for the $\nabla \cdot \mathbf{E} = 4\pi\rho$ equation,

$$\nabla \cdot \mathbf{E} = -\nabla \cdot \left(\nabla V + \frac{1}{c} \partial_t \mathbf{A} \right).$$

Now eliminate $\nabla \cdot \mathbf{E}$ to get

$$\nabla^2 V + \frac{1}{c} \partial_t \underbrace{\nabla \cdot \mathbf{A}}_{\substack{\text{Coulomb} \\ \text{gauge}}} = -4\pi\rho.$$

The equation $\mathbf{B} = \nabla \times \mathbf{A}$ has been used to define $\mathbf{A}$, but $\mathbf{A}$ has not been uniquely set until div $\mathbf{A}$ has also been determined by selecting the function M (Helmholtz's theorem). The div $\mathbf{A}$ relationship can thus be used to specify the appropriate M; picking an M is called *setting the gauge.* Two gauges commonly used (and indicated in the equations above) are (1) the *Lorentz gauge*, for which $\nabla \cdot \mathbf{A} + \partial_t V/c = 0$, and (2) the *Coulomb gauge*, for which $\nabla \cdot \mathbf{A} = 0$.

Maxwell's equations for the Lorentz gauge*

For the Lorentz gauge, MAXWELL'S EQUATIONS become the following:

MAXWELL'S EQUATIONS
FOR THE LORENTZ GAUGE

$$\nabla^2 \mathbf{A} - \frac{1}{c^2} \partial_t^2 \mathbf{A} = -\frac{4\pi}{c} \mathbf{J}$$

$$\nabla^2 V - \frac{1}{c^2} \partial_t^2 V = -4\pi\rho$$

Notice that in regions where there are no charges and no current, these become a pair of wave equations. For this situation, wave equations also describe the ith components E_i and B_i of the electromagnetic field. To see this, take the curl of the equation for $\nabla^2 \mathbf{A}$ for a region without current to get three wave equations (one for each spatial component i) for the magnetic-field result:

WAVE EQUATION
FOR THE MAGNETIC FIELD

$$\nabla^2 B_i - \frac{1}{c^2} \partial_t^2 B_i = 0$$

Note that the propagation velocity of the waves is c, the speed of light.

To find the corresponding equation for the electric field, take the gradient of the $\nabla^2 V$ equation for a region without charge, and use the definition $\nabla V_i = -(E_i + \partial_t A_i/c)$ in the equation for the ith component:

$$\nabla\left(\nabla^2 V - \frac{1}{c}\partial_t^2 V\right)_i = \nabla^2(\nabla V_i) - \frac{1}{c^2}\partial_t^2(\nabla V_i)$$

$$= -\nabla^2\left(E_i + \frac{1}{c}\partial_t A_i\right) + \frac{1}{c^2}\partial_t^2\left(E_i + \frac{1}{c}\partial_t A_i\right)$$

$$= -\nabla^2 E_i + \frac{1}{c}\partial_t^2 E_i - \frac{1}{c}\partial_t\left(\nabla^2 A_i - \frac{1}{c^2}\partial_t^2 A_i\right)$$

$$= 0$$

Thus for the ith component of the electric field, we have the the *wave equation for electric field*:

WAVE EQUATION
FOR THE ELECTRIC FIELD

$$\nabla^2 E_i - \frac{1}{c^2}\partial_t^2 E_i = 0$$

Solution of the wave equations for the electromagnetic field*

For a region free of charge and current, the electromagnetic field is described by the two wave equations

$$\nabla^2 \mathbf{B} = \frac{1}{c^2}\partial_t^2 \mathbf{B} \quad \text{and} \quad \nabla^2 \mathbf{E} = \frac{1}{c^2}\partial_t^2 \mathbf{E}.$$

The plane-wave solutions to these equations are

$$\mathbf{B} = \mathbf{B}_0 e^{i\mathbf{k}\cdot\mathbf{r} - i\omega t} \quad \text{and} \quad \mathbf{E} = \mathbf{E}_0 e^{i\mathbf{k}\cdot\mathbf{r} - i\omega t},$$

where $\mathbf{k}$ is the wave vector that points in the direction of wave propagation with a magnitude of $k = \omega/c$ ($\mathbf{k}$ has units of radians/length), ω is the wave frequency (units of radians/time), $\mathbf{r}$ is the position vector, and $\mathbf{B}_0$ and $\mathbf{E}_0$ are constants. The wave length is $2\pi/k$.

For convenience, orient the z axis so that it points in the $\mathbf{k}$ direction. Then

$$\mathbf{B} = e_2 B_0 e^{ikz-i\omega t} \quad \text{and} \quad \mathbf{E} = e_1 E_0 e^{ikz-i\omega t},$$

where E_0 and B_0 give the wave amplitudes and $\mathbf{e}_1$ and $\mathbf{e}_2$ are of magnitude 1, fixed vectors that specify the direction of **B** and **E**.

These equations describe the electromagnetic field in a region free of charge and current. It turns out that **k**, $\mathbf{e}_1$, and $\mathbf{e}_2$ are all mutually perpendicular, so the magnetic and electric components of the field are always orthogonal to one another, and both are orthogonal to the direction of wave propagation. Furthermore, $B_0 = E_0$ (in the units used here). The goal now is to see why this is true.

Because of the way we have constructed the plane-wave solution, the derivatives of **E** and **B** in the x and y directions vanish. Thus $\nabla \cdot$ reduces to just the ∂_z component. Call the unit vectors in the x, y, and z directions $\mathbf{e}_x$, $\mathbf{e}_y$, and $\mathbf{e}_z$. We know from Maxwell's equations that $\nabla \cdot \mathbf{E} = 0$ in a charge-free region, so

$$\nabla \cdot \mathbf{E} = \mathbf{e}_z \frac{\partial E}{\partial z} = iE_0 \mathbf{e}_z \cdot \mathbf{e}_1 k e^{ikz-i\omega t} = 0.$$

This means that $\mathbf{e}_z \cdot \mathbf{e}_1 = 0$, and because wave propagation is in the direction of $\mathbf{e}_z$, the electric vector $\mathbf{e}_1$ must be orthogonal to the direction of propagation. Since $\nabla \cdot \mathbf{B} = 0$, the same is true for the magnetic component of the field: $\mathbf{e}_z \cdot \mathbf{e}_2 = 0$.

The next job is to show that the electric and magnetic vectors are mutually perpendicular. According to Maxwell's equations, for a charge-free region,

$$\nabla \times \mathbf{B} - \frac{1}{c} \partial_t \mathbf{E} = 0.$$

Substitute the plane-wave solutions into this equation to get

$$\nabla \times \mathbf{e}_2 B_0 e^{ikz-i\omega t} - \frac{1}{c} \partial_t (\mathbf{e}_1 E_0 e^{ikz-i\omega t})$$

$$= \mathbf{e}_z \times \mathbf{e}_2 ikB_0 e^{ikz-i\omega t} + \frac{i\omega}{c} \mathbf{e}_1 E_0 e^{ikz-i\omega t} = 0.$$

This means that

$$ie^{ikz-i\omega t} \underbrace{\left(\mathbf{e}_z \times \mathbf{e}_2 B_0 k + \frac{\omega}{c} \mathbf{e}_1 E_0 \right)}_{\text{vanishes}} = 0,$$

so the indicated quantity must vanish. Recall that $k = \omega/c$. Thus $B_0 = E_0$, and $\mathbf{e}_1 = -\mathbf{e}_z \times \mathbf{e}_2$. The vector $\mathbf{e}_z \times \mathbf{e}_2$ is perpendicular to both $\mathbf{e}_z$ and $\mathbf{e}_2$, so the direction of wave propagation ($\mathbf{e}_z$) and the electric and magnetic vectors ($\mathbf{e}_1$ and $\mathbf{e}_2$) are all mutually perpendicular.

Maxwell's equations for the Coulomb gauge*

For the Coulomb gauge, we have the following versions of Maxwell's equations:

MAXWELL'S EQUATIONS
FOR THE COULOMB GAUGE

$$\nabla^2 V = -4\pi\rho$$

$$\nabla^2 \mathbf{A} - \frac{1}{c^2}\partial_t^2 \mathbf{A} = -\frac{4\pi \mathbf{J}_\mathrm{T}}{c}$$

Here $\mathbf{J}_\mathrm{T}$ is the transverse component of the current, defined below. The first equation follows at once from the definition of the Coulomb gauge, but seeing where this last equation comes from takes a few steps. The starting place is the equation for $\nabla^2\mathbf{A}$:

$$\nabla^2 \mathbf{A} - \frac{1}{c^2}\partial_t^2 \mathbf{A} - \frac{1}{c}\partial_t \nabla V = -\frac{4\pi \mathbf{J}}{c}$$

The problem is to get rid of the term involving V. To do this, start by writing $\mathbf{J}$ as a sum of its transverse part $\mathbf{J}_\mathrm{T}$ (for which $\nabla \cdot \mathbf{J}_\mathrm{T} = 0$, according to Helmholtz's theorem) and its longitudinal part $\mathbf{J}_\mathrm{L}$ (for which $\nabla \times \mathbf{J}_\mathrm{L} = 0$):

$$\mathbf{J} = \mathbf{J}_\mathrm{T} + \mathbf{J}_\mathrm{L}$$

According to Helmholtz, the longitudinal part $\mathbf{J}_\mathrm{L}$ can be written as

$$\mathbf{J}_\mathrm{L} = -\nabla \frac{1}{4\pi} \int \frac{\nabla' \cdot \mathbf{J}}{|\mathbf{x} - \mathbf{x}'|} d^3x'.$$

Now use the continuity equation to eliminate $\nabla \cdot \mathbf{J}$. The resulting integral is the definition of the potential V (because, with the Coulomb gauge, V follows Poisson's equation and thus is just like the potential for electrostatics). This definition gives

$$\mathbf{J}_\mathrm{L} = -\partial_t \frac{\nabla V}{c}.$$

This last equation (together with $\mathbf{J} = \mathbf{J}_T + \mathbf{J}_L$) can be used to eliminate the bothersome $\partial_t \nabla V/c$ in the starting equation to give the desired result.

Force on moving charges*

Knowing the force on a moving charge is important in many instances. To find this force, I turn to an experiment that provides the relationship between the magnetic field and the force experienced by a current-carrying wire; this information will lead to the Lorentz relation for electromagnetic-field forces on moving charges. The force $\mathbf{F}$ experienced by a loop of wire with current I in the presence of magnetic field $\mathbf{B}$ is

$$\mathbf{F} = \frac{I}{c} \oint d\mathbf{s} \times \mathbf{B}.$$

This equation specifies the force a current-carrying wire experiences in a magnetostatic field and thus connects magnetostatics defined by permanent magnets with a magnetic field derived from current flow. For current flowing in a volume rather than in a fine wire, the last equation becomes

$$\mathbf{F} = \frac{1}{c} \int \mathbf{J} \times \mathbf{B}\, d^3x$$

(substitute $\mathbf{J}$ for $I_0\, d\mathbf{s}$, and use the fact that magnetic fields add linearly). And for an element of wire $d\mathbf{s}$ with current I_0, it becomes

$$d\mathbf{F} = \frac{I_0}{c} d\mathbf{s} \times \mathbf{B}.$$

For a moving charge element dq, the force, according to the preceding equation, should be

$$d\mathbf{F} = \frac{dq}{c} \frac{d\mathbf{s}}{dt} \times \mathbf{B} = \frac{dq}{c} (\mathbf{v} \times \mathbf{B}),$$

where $\mathbf{v} = d\mathbf{s}/dt$ is the velocity of the charge element. If an electric field $\mathbf{E}$ is also present, the force on the charge element would have the additional component $dq\,\mathbf{E}$. Now integrate over all of the charge elements moving in a tiny region with volume V to get the total force $\mathbf{f}$ experienced by a local charge density $\rho = (1/V) \int_V dq$. This gives the *Lorentz force on moving charges*:

LORENTZ FORCE ON MOVING CHARGES

$$\mathbf{f} = \rho\left(\mathbf{E} + \frac{1}{c}\mathbf{v} \times \mathbf{B}\right)$$

In the chapter on classical mechanics (page 65), I noted that the Lorentz force, although not conservative, can be gotten from a velocity-dependent potential that still meets the requirements for the Euler-Lagrange equation to hold. Specifically, if the force can be calculated from the potential U according to the relation

$$\text{force} = \frac{d}{dt}\left(\frac{\partial U}{\partial q_i}\right) - \frac{\partial U}{\partial q_i},$$

then a particle acting under this force will still obey the Euler-Lagrange equation. If the potential is defined by

$$U = q\left(V - \frac{1}{c}\mathbf{A}\cdot\mathbf{v}\right),$$

then the Lorentz force $\mathbf{f}$ on a single particle with charge q,

$$\mathbf{f} = q(\mathbf{E} + \mathbf{v} \times \mathbf{B}/c),$$

will result from the operations to derive $\mathbf{f}$ from U that are needed to make the Euler-Lagrange equations valid. To see this, start by writing $\mathbf{f}$ in terms of the vector and scalar potentials:

$$\mathbf{f} = q\left(-\nabla V - \frac{1}{c}\partial_t\mathbf{A} + \frac{1}{c}\mathbf{v} \times \nabla \times \mathbf{A}\right)$$

Use the vector relationship $\mathbf{a} \times \mathbf{b} \times \mathbf{c} = (\mathbf{a}\cdot\mathbf{c})\mathbf{b} - (\mathbf{a}\cdot\mathbf{b})\mathbf{c}$ for the last term to get

$$\mathbf{v} \times \nabla \times \mathbf{A} = \nabla(\mathbf{v}\cdot\mathbf{A}) - (\nabla\cdot\mathbf{v})\mathbf{A} = \nabla(\mathbf{v}\cdot\mathbf{A}).$$

The term containing $\nabla\cdot\mathbf{v}$ vanishes because $\mathbf{v}$ is proportional to current $\mathbf{j}$ and $\operatorname{div}\mathbf{j}$ vanishes for a single moving charge. This means that, in terms of potentials,

$$\mathbf{f} = q\left(-\nabla V - \frac{1}{c}\partial_t\mathbf{A} + \frac{1}{c}\nabla(\mathbf{v}\cdot\mathbf{A})\right).$$

We now have to show that we get the same result with the Euler-Lagrange prescription starting from U. The gradient of U at once gives

$$\nabla U = q\left(\nabla V - \frac{1}{c}\nabla(\mathbf{v}\cdot\mathbf{A})\right),$$

and the $d(\partial U/\partial\dot{q})/dt$ term, in three dimensions, gives (with nonstandard but obvious notation)

$$\frac{d}{dt}\left(\frac{\partial U}{\partial \mathbf{v}}\right) = \frac{1}{c}\partial_t\mathbf{A}.$$

Together, then, these two components give the Lorentz force on a moving charged particle.

4 Quantum Mechanics

Unlike classical mechanics, or even classical electricity and magnetism, quantum mechanics does not lend itself to a simple "derivation" from the results of a few experiments. The reason is that the theory evolved over a quarter of a century under the pressure of many different sorts of experimental observations and hand in hand with the mathematics necessary for its modern formulation. Hence the cleanest way to develop quantum mechanics is to start with a set of postulates, develop the formalism, and compare predictions with experiments. But this approach is unmotivated and therefore tends to be opaque. The inductive alternative—building the theory bit by bit either historically or through a series of selected experimental observations—gives better intuitions about what it all means and dispels some of the rabbit-out-of-the-hat feeling, but it is cumbersome for a brief summary of the subject and tends to obscure the form of the theory. This presentation is somewhat in between these alternatives. The development here provides some motivation but is not really a derivation of the same sort that led, for example, to the equations of classical mechanics from Newton's law. This choice is selected for two reasons: the first is to give some feeling of reasonableness, and the second to present the Feynman formulation of quantum mechanics that is the one needed for field theory (chapter 7) and quantum statistical mechanics (page 164).

The first step is to treat the movement of a single particle in one dimension from one place to another according to the Feynman view of quantum mechanics. Then we will derive the usual Schrödinger equation. Following this, we will consider how momentum, in addition to the particle's position, is treated. This will then lead to a general treatment of measurement and the notion of operators as the representation of dynamic variables. A development of the uncertainty principle will lead to the notation of commutators and then to a comparison of the Heisenberg and Schrödinger pictures. In the final section I treat the problem of time-varying forcing to develop some concepts that will be used later in field theory.

Fundamental Assumptions

The starting place is the observation that very small particles—photons or electrons, for example—are simultaneously particles and waves, and that their behavior as particles is probabilistic. For example, if extremely low intensity light is sent through a diffraction grating toward an appropriate detector, individual photons are found to arrive at the detector with a position that is probabilistically determined. At the same time, the spatial pattern of arrivals is what one would predict for waves passing through the diffraction grating. Thus a single tiny photon (know to be single because just one was detected) has its behavior determined by a considerably extended diffraction grating: a particle is somehow represented by a wave that interacts with itself constructively and destructively to give probabilistic photon arrivals that build into the diffraction pattern when enough photons have gone through the diffraction grating.

We need a theory, then, in which particles can interact with their surroundings and with themselves (in a single-particle diffraction experiment, for example) in a wavelike way to produce interference patterns but at the same time can be observed to arrive probabilistically as particles at a particular location and time.

The state of a particle will be specified by a function that gives the *amplitude* of the particle's wave at a particular place x and time t. We denote the amplitude with $\psi(x,t)$, which is a complex number for easy representation of the particle's wave nature: we need waves to get the interference of single particles observed in diffraction experiments. This representation of amplitudes as complex numbers provides a convenient mathematical description of waves, but it is actually a postulate of the theory. The probability $P(x,t)$ of finding a particle at x and t is then assumed to be $|\psi(x,t)|^2$, just as the magnitude of an electric or acoustic wave would be found in that way. This is a *fundamental assumption of quantum mechanics*:

FUNDAMENTAL ASSUMPTION
OF QUANTUM MECHANICS

$$P(x,t) = |\psi|^2$$

So

$$\int \psi^*(x,t)\psi(x,t)\,dx = 1$$

for all t.

The job is to find an equation for ψ. But the solution of any equation for ψ must always approach the classical trajectory as we consider successive systems that make the transition from microscopic to macroscopic: this is the important *correspondence principle*. In classical mechanics, the path of a particle is always the one that minimizes THE ACTION $S = \int L(x, \dot{x}, t)\, dt$ for the Lagrangian L (see the discussion of HAMILTON'S PRINCIPLE on page 62). The particle acts as if it knows the value of L and follows the path that makes it a minimum. In microphysics, a single particle following its individual trajectory does not conform to this classical path (think of the diffraction experiment), but as progressively more massive particles are studied, their paths are more and more likely to be the classical one. This observation suggests that the Lagrangian determines the amplitude for the particle's wave so that all paths have an amplitude related to the action for that particular path, and that macroscopic systems then have a totally dominant amplitude for the classical path.

Imagine a particle going from $\xi(\tau)$ to $x(t)$, and call the amplitude for this event $K(x, t; \xi, \tau)$. The corresponding conditional transition probability density would be KK^*. Then according to the notions in the preceding paragraph, the total amplitude to end up at the final place (x, t) should be the sum of the amplitudes for all the ways to get there. For any one path specific path $x(t')$, the amplitude is assumed to be proportional to

$$\exp\left(\frac{i}{\hbar} S[x(t')]\right) = \exp\left(\frac{i}{\hbar} \int_\tau^t L(x, \dot{x}, t')\, dt'\right),$$

where S is the action associated with that particular path and $2\pi\hbar$ is Planck's constant. For two paths, $y_0(t')$ and $y_1(t')$, both of which start at $\xi(\tau)$ and end at $x(t)$, the transition amplitude would be proportional to

$$\exp\left(\frac{i}{\hbar} S[y_0(t')]\right) + \exp\left(\frac{i}{\hbar} S[y_1(t')]\right).$$

Suppose next—this is *definitely not* true—that a particle can move from ξ to x only over a countable number of paths: $y_0, y_1, \ldots$. For this hypothetical case, the transition amplitude would be

$$K(x, t; \xi, \tau) = \frac{1}{\theta} \sum_k \exp\left(\frac{i}{\hbar} S[y_k(t')]\right).$$

(The factor θ is included to normalize the sum so that KK^* is a probability.) Note that since the total transition probability depends on contributions from all of the possible paths, interactions of the particle with itself and with its surrounding are included and account for such effects as the single-particle diffraction pattern described earlier. The problem, of course, is that the number of paths is not countable, so the sum given above is not correct. How are we to calculate the sum of all weights for every conceivable path?

To make this calculation, we must, according to the preceding assumption, find the action for every possible path. Take a representative path $x(t')$ that starts at ξ (at time τ) and ends at x (at time t), and approximate this path with N straight-line segments that connect the points $\xi, x_1, x_2, \ldots, x_{N-1}, x; x_k \equiv x(t_k)$, where $t_k = k\Delta t$, and $\Delta t \equiv (t - \tau)/N$. Call this path $X_N(\xi, x_1, \ldots, x_{N-1}, x)$, and note that it is a function of $N - 1$ arguments, the points x_k that the straight-line segments connect. All possible paths can be represented in this way, so any other path too would be given by X_N but with different arguments (to specify the points through which that path actually passes). The action associated with a path X_N is $S[X_N]$, and the action too is a function of the $N - 1$ variables needed to specify approximately any particular path. Also, for any specific path, $S[X_N] \to S[x(t')]$ as $N \to \infty$. Now as each of the x_k is varied, we will scan through all of the possible paths. The transition amplitude is the sum of the weights associated with all of these paths, so

$$K(x, t; \xi, \tau) \approx \underbrace{\int \cdots \int}_{N-1} \frac{dx_1 dx_2 \cdots dx_{N-1}}{\theta} \exp\left(\frac{i}{\hbar} S[X_N]\right).$$

This integral sums over all paths by varying $x_1, x_2, \ldots, x_{N-1}$ over all of their possible values in all combinations; the factor θ normalizes the integral so that KK^* can be interpreted as a probability.

As $N \to \infty$, the calculation of the transition amplitude will become exact, and the $(N - 1)$-fold integral, by definition, approaches the *key-path integral*:

KEY-PATH INTEGRAL

$$K(x, t; \xi, \tau) = \int \exp\left(\frac{i}{\hbar} \int_{\tau}^{t} L(x, \dot{x}, t')\, dt'\right) \mathscr{D}x(t)$$

Here $2\pi\hbar$ is Planck's constant. (Refer to the section "Functional Calculus" in chapter 1 [pages 40–44] for a description of path integrals.) This relationship gives, as it should, a complex amplitude for K, but it also predicts the classical path as the most likely one for any particle, and overwhelmingly so as the particle mass, and hence the action, gets large in comparison with the scale-setting constant $\hbar$. This prediction that a sufficiently massive particle follows the classical path arises because the particular path that makes the action

$$S = \int_{\tau}^{t} L\,dt'$$

vanish dominates the integral for K (use a steepest-descent or stationary-phase argument) and is therefore most probable. But according to HAMILTON'S PRINCIPLE, the classical path is just the one that causes the action to vanish, and this is what is needed for that particular path (the classical path) to dominate the path integral.

A power-series approach shows how the equation for K might arise. The amplitude is assumed to be a complex number and should be some (complex-valued) function F of the action S, which in turn is a (real-valued) functional of the particle's path. Thus, to first order,

$$K(x,t;\xi,\tau) = \int F(S)\,\mathscr{D}x = \int \exp(i\ln(F(S)))\,\mathscr{D}x \approx \int \exp\left(\frac{i}{\hbar}S\right)\mathscr{D}x,$$

where $\hbar$ is the coefficient in the power-series expansion of $\ln(F)$. Note that writing $F(S) = \exp(i\ln F)$ provides a slowly varying function for the power-series expansion and that the i is necessary to make amplitude K complex, as it has been assumed to be. The zeroth-order term does not appear, because it would make a constant contribution to the transition amplitude for all trajectories. Moreover, higher-order terms must vanish, because otherwise the particle's average path would not follow Hamilton's principle.

The transition amplitude K behaves like a conditional probability. That is,

$$K(x,t;\xi,\tau) = \int K(x,t;y,t')K(y,t';\xi,\tau)\,dy$$

because this is a linear theory, so amplitudes are additive, and to get from ξ to x, the particle must go through some $y(t')$. Thus this preceding equation gives a representation of the sum of all the ways to go from the starting $\xi(\tau)$ to the final $x(t)$. The equation arises as follows. By representing the action as the sum of two integrals, we can write the amplitude $K(x, t; \xi, \tau)$ as

$$K(x, t; \xi, \tau) = \int \mathscr{D}x \exp\left(\frac{i}{\hbar}\int_{\tau}^{t'} L\, du + \frac{i}{\hbar}\int_{t'}^{t} L\, du\right),$$

where the system must pass through some intermediate position y at time t'. Recall that this functional integral is really the limit of an N-fold integral, where $N \to \infty$, and that the volume element $\mathscr{D}x$ is the limit of the product $\prod dx_k$. The idea is to break down the integral for K into three parts: one for all x up to the intermediate $y(t')$, one that uses the volume element $dx_k = dy(t')$, and one for the rest of the xs. For the τ to t' integration, which is performed first, the t' to t integral is a constant because we only consider paths from ξ to y. Thus

$$K(x, t'; \xi, \tau) = \int \exp\left(\frac{i}{\hbar}\int_{t'}^{t} L\, du\right) K(y, t'; \xi, \tau)\, \mathscr{D}x\, dy(t').$$

Next carry out the integration over paths from $y(t')$ to $x(t)$. Since we must finally integrate over all possible $y(t)$, this gives the desired equation:

$$K(x, t; \xi, \tau) = \int K(x, t; y, t') K(y, t'; \xi, \tau)\, dy$$

Let $\psi(x, t)$ be the unconditional amplitude for the particle being at $x(t)$. Then the preceding equation becomes

$$\psi(x, t) = \int K(x, t; \xi, \tau)\psi(\xi, \tau)\, d\xi.$$

This last expression provides the required specification for the amplitude (or *wave function*, in the usual terminology) for a single particle.

The specific form for the transition amplitude for a free particle will be important for the development that follows. This amplitude is found by observing that the Lagrangian for a free particle of mass m is $L(t) = m\dot{x}^2/2$ (kinetic energy minus potential energy, and the potential energy for a free particle vanishes). So we have K_0, the free-particle transition amplitude:

$$K_0(x, t; \xi, \tau) = \int \exp\left(\frac{i}{\hbar}\int_{\tau}^{t} m\dot{x}^2(t')/2\, dt'\right) \mathscr{D}x;$$

the path integral is evaluated over all paths that start at $\xi(\tau)$ and end at

$x(t)$. This is a Gaussian integral that can be evaluated to yield the *free-particle transition amplitude*:

FREE-PARTICLE TRANSITION AMPLITUDE

$$K_0(x, t; \xi, \tau) = \left(\frac{2\pi i\hbar(t - \tau)}{m}\right)^{-1/2} \exp\left(\frac{im(x - \xi)^2}{2\hbar(t - \tau)}\right)$$

(For the evaluation of Gaussian integrals, see the subsection "Functional Integration" in chapter 1, pages 40–44.)

Schrödinger's Equation

The standard and, for many practical purposes, most convenient formulation of quantum mechanics makes use of the familiar Schrödinger equation. This equation is a consequence of the path-integral description just presented, and this section establishes how the Feynman transition amplitude gives the standard differential equation for the wave function. The strategy is to look at the transition amplitude K at very short times and then to use a power-series expansion of the Feynman expression for the particle's wave function ψ.

Suppose that a particle with a mass m is moving in a potential V from a starting position $x + \xi$ at time t and then ends up at $x(t + \tau)$. The transition amplitude for this event is

$$\begin{aligned}
K(x, t + \tau; x + \xi, t) &= \int \mathscr{D}x \exp\left(\frac{i}{\hbar}\int_t^{t+\tau} L(x, \dot{x}, t')\,dt'\right) \\
&= \int \mathscr{D}x \exp\left(\frac{i}{\hbar}\int_t^{t+\tau} \left(\frac{m\dot{x}^2}{2} - V(x, t')\right)dt'\right) \\
&\approx \int \mathscr{D}x \exp\left(\frac{i\tau}{\hbar}\left(\frac{m\xi^2}{2\tau^2} - V(x, t)\right)\right) \\
&= M \exp\left(\frac{i\tau}{\hbar}\left(\frac{m\xi^2}{2\tau^2} - V(x, t)\right)\right).
\end{aligned}$$

(Remember that $L = mv^2/2 - V$, and notice that $\dot{x}^2 = (\xi/\tau)^2$.) Here

$$M \equiv \int \mathscr{D}x$$

must be determined later; this expression gives the transition amplitude K for short time intervals τ.

Next write the equation for the wave function ψ:

$$\psi(x, t + \tau) = \int d\xi\, K(x, t + \tau; x + \xi, t)\psi(x + \xi, t)$$

And then expand to first order in τ and second order in ξ:

$$\begin{aligned}
&\psi(x,t) + \tau\partial_t\psi(x,t) \\
&\quad = \int M \exp\left(\frac{i\tau}{\hbar}\left(\frac{m\xi^2}{2\tau^2} - V(x,t)\right)\right)\left[\psi(x,t) + \xi\partial_x\psi + \frac{\xi^2}{2}\partial_x^2\psi\right]d\xi \\
&\quad \approx \int \exp\left(\frac{i}{\hbar}\frac{m\xi^2}{2\tau}\right)\left(1 - \frac{i\tau}{\hbar}V\right)\left[\psi(x,t) + \xi\partial_x\psi + \frac{\xi^2}{2}\partial_x^2\psi\right]d\xi \\
&\quad = \left(1 - \frac{i\tau}{\hbar}V\right)[I_1\psi + I_2\partial_x\psi + I_3\tfrac{1}{2}\partial_x^2\psi]
\end{aligned}$$

The three integrals I_1, I_2, and I_3 that appear in the last step are defined as follows:

$$I_1 = M\int d\xi \exp\left(\frac{im\xi^2}{2\hbar\tau}\right)$$

$$I_2 = M\int d\xi \exp\left(\frac{im\xi^2}{2\hbar\tau}\right)\xi$$

$$I_3 = M\int d\xi \exp\left(\frac{im\xi^2}{2\hbar\tau}\right)\xi^2$$

The reason that the expansion is to first order in τ and second order in ξ is that these quantities are present in the equation to these powers. To evaluate the three integrals, one must know that

$$\int_{-\infty}^{\infty} dx\, e^{iax^2} = \sqrt{\pi/\alpha}\, e^{i\pi/4} = \sqrt{i\pi/\alpha},$$

where the last step uses the fact that $(e^{i\pi/4})^2 = e^{i\pi/2} = \cos\pi/2 + i\sin\pi/2 = i$.

The easy integral to start with is the second one. $I_2 = 0$ because the integrand is an odd function (the even exponential function is multiplied by the odd function ξ) and thus the integral vanishes when it is evaluated from $-\infty$ to ∞. Next use the identification $\alpha = m/2\hbar\tau$ in the definite integral above to evaluate I_1:

$$I_1 = M\sqrt{\frac{2\pi i\hbar\tau}{m}}$$

In order for the equation for ψ to hold in the limit $\tau \to 0$, the integral I_1 must approach 1. Thus

$$M = \sqrt{\frac{m}{2\pi i\hbar\tau}},$$

and

$$I_3 = \sqrt{\frac{m}{2\pi i\hbar\tau}}\int d\xi \exp\left(\frac{im\xi^3}{2\hbar\tau}\right)\xi^2 = \frac{i\hbar\tau}{m}.$$

This last relation can be found by taking $\partial I_1/\partial\alpha$ with $\alpha = (2\hbar\tau)/m$ and the value of M inserted; the derivative operation brings ξ^2 down so that I_3 can be expressed in terms of the value for I_1.

So the equation for ψ becomes, to first order in τ,

$$\begin{aligned}\psi + \tau\partial_r\psi &= \left(1 - \frac{i\tau}{\hbar}V\right)\left(\psi + \frac{i\hbar\tau}{2m}\partial_x^2\psi\right) \\ &\approx \psi + \frac{i\hbar\tau}{2m}\partial_x^2\psi - \frac{i\tau}{\hbar}V\psi.\end{aligned}$$

Thus, for small τ (and for a particle that moves in only one dimension x in a potential V), this is *Schrödinger's equation*:

SCHRÖDINGER'S EQUATION

$$-\frac{\hbar}{i}\frac{\partial\psi}{\partial t} = -\frac{\hbar^2}{2m}\frac{\partial^2\psi}{\partial x^2} + V\psi$$

This equation can be written as

$$-\frac{\hbar}{i}\partial_t\psi = \mathbf{H}\psi,$$

with the operator $\mathbf{H}$ (for a particle of mass m in potential V) defined as

$$\mathbf{H} = -\frac{\hbar^2}{2m}\partial_x^2 + V.$$

$\mathbf{H}$ is known as the *Hamiltonian.* In three dimensions, ∇^2 replaces the ∂_x^2 in Schrödinger's equation displayed above.

Because we have the relation

$$\psi(x,t) = \int d\xi\, K(x,t;\xi,\tau)\psi(\xi,\tau),$$

the transition amplitude $K(x,t;\xi,\tau)$ must be the Green's function (or impulse response) for the Schrödinger equation (see the discussion of Green's functions on page 22). That is, K satisfies the equation

$$\frac{\hbar}{i}\frac{\partial K}{\partial t} + \mathbf{H}K(x,t;\xi,\tau) = \delta(t-\tau)\delta(x-\xi).$$

Here the delta function in the spatial coordinates is unitless, but the delta function in the temporal coordinate must have assigned to it the units of energy because these are the units of the left side.

Up to this point we have focused, for clarity and simplicity, on systems of a single particle so the description has mostly depended on only a single spatial variable, or at most three. The same formalism also applies, however, to a much larger class of systems. Picture a collection of particles with positions $\mathbf{x}_1, \mathbf{x}_2, \ldots$. For such a system, the Lagrangian is, as before, the difference between kinetic and potential energy, except that now the kinetic energy is a sum of the individual-particle contributions and the potential energy in general depends simultaneously on all of the particle positions (because they all may interact with one another). A transition amplitude (which now depends on all of the coordinates) is defined, just as before, by a multidimensional functional integral, and the manipulations that led to Schrödinger's equation are unchanged except that the amplitude $\psi(\mathbf{x}_1, \mathbf{x}_2, \ldots)$ is now a function of all the coordinates and the Hamiltonian $\mathbf{H}$ is correspondingly more complicated. Indeed, the formalism can be extended to more general systems in which the Lagrangian depends on coordinates other than the positions of individual particles. In every case, however, the same principle for the path integral provides the transition amplitude. We will continue to use the simple example of a single particle moving in a one-dimensional space, but all of the results to be described generalize at once to the multidimensional situation.

Next Steps

The Schrödinger equation was written above as an operator equation:

$$-\frac{\hbar}{i}\partial_t\psi = \mathbf{H}\psi,$$

where **H** is the Hamiltonian operator defined on page 112. In standard treatments of quantum mechanics, operators play a central role. Every observable—like position, momentum, energy—is associated with its own operator, and that operator can be used to calculate the average value of the observable, the quantity that would normally be measured in an experiment. For example, the operator that corresponds to the observable *momentum* in the one-dimensional case is, we will see, $(\hbar/i)(\partial/\partial x)$, and for a particle with a wave function $\psi(x)$, the average value for the particle's momentum is

$$\int dx\,\psi^*(x)\left(\frac{\hbar}{i}\right)\frac{\partial\psi(x)}{\partial x}.$$

This connection between observable and operator comes naturally out of the Feynman approach, but only after a few steps. The goal now is to identify the operators associated with observables. To start, we will have to answer the question, What is the transition amplitude for momentum (analogous to the transition amplitude K for position that we have used to this point)? Answering this question will provide the model for a definition of operators, which will finally enable us to treat the connection between observable and operator in general.

Momentum Representation

The wave function ψ described above contains all of the relevant information about the particle's state and permits the probability of a particle's position to be calculated. Suppose, however, that the probability of the momentum rather than the position is needed. The goal of this section is to describe how the momentum amplitude $\phi(p)$ is found from the wave function $\psi(x)$.

The transform from spatial representation to momentum representation will turn out to be simply a Fourier transform. To see how this arises, picture the way a particle's momentum might be measured. The momentum $p = m\dot{x}$, so if a particle does not interact with its surroundings over the interval 0 to T (take the initial position of the particle to be approximately 0), the (unchanging) momentum is just $m(x(T)/T)$. The amplitude for a particle to arrive at $x(T)$ from a starting location near 0 (remember that there are no interactions, so the particle is free) is

$$\psi(x, T) = \int K_0(x, t; \xi, 0)\psi(\xi)\, d\xi,$$

and for a free particle (use the FREE-PARTICLE TRANSITION AMPLITUDE),

$$\begin{aligned}\psi(x, T) &= \left(\frac{m}{2\pi i\hbar T}\right)^{1/2} \int d\xi \exp\left(\frac{im}{2\hbar T}(x - \xi)^2\right)\psi(\xi)\\ &= \left(\frac{m}{2\pi i\hbar T}\right)^{1/2} \exp\left(\frac{-imx^2}{2\hbar T}\right) \int d\xi \exp\left(\frac{im}{2\hbar T}(\xi^2 - 2x\xi)\right)\psi(\xi).\end{aligned}$$

The probability density for finding the particle at x is given by

$$|\psi(x, T)|^2 = \frac{m}{2\pi T\hbar}\left|\int d\xi \exp\left(\frac{im\xi^2}{2T\hbar} - \frac{i\xi}{\hbar}\left(\frac{mx}{T}\right)^2\right)\right|^2,$$

and because the particle is free, this also gives the probability of finding a momentum of $p = mx/T$. Use the relation $p = mx/T$, so that $dp = m(dx/T)$, to write the expression for $P(p)\, dp$, the probability density for momentum p, as

$$|\psi(x, t)|^2\, dx = P(p)\, dp = \frac{dp}{2\pi\hbar}\left|\int d\xi\, \psi(\xi, 0) \exp\left(\frac{im\xi^2}{2\hbar T} - \frac{ip\xi}{\hbar}\right)\right|^2.$$

Remember that $\psi(\xi)$ vanishes except near 0 (because the particle starts near 0). As $T \to \infty$, the integral approaches the true momentum (because the uncertainty in the initial position is progressively less important), and it becomes

$$P(p)\, dp = \frac{dp}{2\pi\hbar}\left|\int d\xi\, \psi(\xi, 0) \exp\left(\frac{ip\xi}{\hbar}\right)\right|^2 = \frac{dp}{2\pi\hbar}|\phi(p)|^2.$$

The amplitude for momentum, then, is just

$$\phi(p) = \int dx\, \psi(x) \exp\left(\frac{-ipx}{\hbar}\right).$$

Note that this equation is just a Fourier transform of ψ with the Fourier transform variable $p/\hbar$. This means that ψ can be calculated from ϕ by an inverse Fourier transform with the transform variable $\xi/\hbar$. That is,

$$\psi(x,t) = \int \exp((i/\hbar)px)\phi(p,t)\frac{dp}{2\pi\hbar}.$$

Thus a Fourier transform is used to switch between the position and momentum representations.

Operators

If all of quantum mechanics involved only simple generalizations of familiar classical quantities like position, momentum, and energy, the machinery developed in the preceding sections would be mostly adequate. In fact, however, a complete treatment of microphysics turns out to involve properties of particles, like *spin*, without classical counterparts. To generalize quantum mechanics so that it can deal with these properties of matter that lack strict classical analogs, a more general formulation of certain concepts is required. One of the most important of these is the notion of quantum-mechanical operators, and their relation to measurable quantities.

In the preceding section I developed the mathematical operations that permit the momentum and position amplitudes to be calculated from one another. In that discussion I connected an observable property of a particle (momentum, for example) with a measurement (an experiment in which momentum can be determined) that is associated with a mathematical operator (in this case, a Fourier transform). The goal of this section is first to generalize this process and then to explicitly identify operators for position, momentum, and energy usually used in quantum mechanics. Note that this section assumes a familiarity with the mathematics of linear operators on Hilbert spaces usually employed in quantum mechanics (see the subsection "Hilbert space" starting on page 19). We start with the notion of measurement and then proceed to a calculation of the average value of a property of a particle (like its momentum). The first step is to ask how a general description of measurement can be achieved.

To determine the value a of some property A, the measuring equipment must be arranged to find the value of the variable associated with that property (picture the measurement of momentum described earlier, where A was the momentum of the particle and a its value). If a particle starts out with amplitude $\psi(x)$, the amplitude $\phi(a)$ for finding the value a of property A should be the following, as before, for the transition amplitude $K_a(x)$:

MEASUREMENT OF PROPERTY A

$$\phi(a) = \int K_a(x)\psi(x)\,dx$$

Specifically, $K_a(x)$ gives the amplitude for finding the value a of a property A, given that the particle starts at position x; thus if a particle enters the measuring apparatus at x, the amplitude for finding a is K_a. And $|\phi(a)|^2$ is the probability that the measuring apparatus will report a particle with value a for A when it starts with a wave function $\psi(x)$. How K_a might be calculated, that is, related to the appropriate Lagrangian L, has to be determined from the nature of each A. Earlier I gave a specific example of how this process can be carried out to measure momentum. Note that the equation for the measurement of property A equation defines the transform from the spatial representation to the representation A (e.g., momentum), no one representation is sacred; one could start just as well with, say, the momentum representation and define the operator that gives the amplitude for the property A, just as we went from the momentum representation to the position representation in the preceding section.

A consideration of how the average value of A is measured will lead to the definition of operators in the Feynman formulation of quantum mechanics. The average of A is

$$\begin{aligned}\langle a \rangle &\equiv \int da\,|\phi(a)|^2 a \\ &= \int da\,a \int\!\!\int dx dx'\, K_a^*(x)\psi^*(x) K_a(x')\psi(x').\end{aligned}$$

So

$$\langle a \rangle = \int dx \underbrace{\int dx'\, A(x,x')\psi(x')}_{\mathbf{A}\psi}\psi^*(x),$$

where

$$A(x, x') = \int da\, K_a(x) K_a^*(x') a$$

defines the kernel of a linear operator **A** that acts on ψ,

$$A\psi = \int dx'\, A(x, x')\psi(x'),$$

and corresponds to property A. Thus the property A whose average value is $\langle a \rangle$ becomes associated with the linear operator **A**. Every measurable quantity (property) is associated with a linear operator in this way. Thus for property A, we may write the *average value of A*:

AVERAGE VALUE OF A

$$\langle a \rangle = \int \psi^* \mathbf{A} \psi \, dx$$

Because

$$A^*(x, x') = \left(\int da\, K_a(x) K_a^*(x') a \right)^* = A(x', x),$$

the operator **A**, by the way it is constructed, is Hermitian (see page 20):

$$\int \psi^* \mathbf{A} \psi \, dx = \int \psi \mathbf{A}^* \psi^* \, dx$$

Hermitian operators have real eigenvalues a_n and eigenvectors u_n:

$$\mathbf{A} u_n = a_n u_n$$

Note that the spectrum of eigenvalues might be continuous, discrete, or a mixture of both. If the spectrum of eigenvalues is continuous, then a and u are functions of a continuous variable n, rather than having an integer subscript n if the spectrum is discrete. These eigenvectors (assuming they form a complete set) can be used to represent the wave function. If the spectrum of eigenvalues is continuous, then the state of the system can be represented by

$$\psi = \int dn\, w_n u_n,$$

and if the spectrum is discrete, it can be represented by

$$\psi = \sum w_n u_n,$$

for some set of weights w_n that are determined from the basis vectors u_n with the dot products given by

$$w_n = \int dx\, \psi^*(x) u_n(x)$$

(see page 19). According to these relations, if the system is in the state ψ, the average value of A (for a discrete eigenvalue spectrum) is given by

$$\int \psi^* \mathbf{A} \psi \, dx = \int \left(\sum w_n u_n(x) \right)^* \mathbf{A} \left(\sum w_n u_n(x) \right) dx$$
$$= \sum a_n |w_n|^2.$$

The last step used $\mathbf{A} u_n = a_n u_n$ and the fact that the basis vectors have been chosen to be orthonormal ($\int u_j^* u_k \, dx = \delta_{jk}$). If the operator has a continuous spectrum of eigenvalues, integrals rather than sums result.

Because of these last relationships, it follows that the only possible values of a that can be measured are one of the a_n or weighted combinations of them. If the system is prepared so that it is in a *pure state*, which means that $\psi = w_k u_k$ for just one particular discrete eigenvector u_k, then the any measurement of this property must always give the value a_k (the eigenvalue that corresponds to the eigenvector u_k). This fact was the historical starting place for quantum theory because the measurement of certain properties were found to have only discrete values, that is, to be *quantized*.

An important example of eigenfunctions and eigenvalues is provided by solutions to SCHRÖDINGER'S EQUATION for the special case in which the Hamiltonian is not explicitly dependent on time. For this situation, we can seek solutions of Schrödinger's equation that are separable:

$$\psi(x, t) = f(t)\phi(x),$$

where $f(t)$ is a function of time but not of position and $\phi(x)$ is a function of position but not of time. Substitute this into Schrödinger's equation to give

$$-\frac{\hbar}{i} \frac{\partial \psi(x,t)}{\partial t} = -\frac{\hbar}{i} \phi(x) \frac{df(t)}{dt} = \mathbf{H}\psi(x,t) = f(t)\mathbf{H}\phi(x),$$

which rearranges to

$$-\frac{\hbar}{i}\frac{df/dt}{f(t)} = \frac{\mathbf{H}\phi}{\phi} = E.$$

Since the left side is a function of time (but not of position) and the right side is a function of position (but not of time), both sides must be equal to a constant, which we call E. This gives a pair of equations:

$$-\frac{\hbar}{i}\frac{df}{dt} = Ef(t)$$

$$\mathbf{H}\phi(x) = E\phi(x)$$

This last relation defines an eigenvalue problem, so the equation has eigenvalues E_n and eigenfunctions $\phi_n(x)$ that define, as noted earlier, a complete orthonormal basis. The time-dependent equation is easily solved (up to a multiplicative constant that is normalized away in the wave function) to give

$$f(t) = \exp\left(-\frac{i}{\hbar}E_n t\right).$$

Thus the wave functions have an oscillatory form,

$$\psi_n(x,t) = f(t)\phi_n(x) = \exp\left(-\frac{i}{\hbar}E_n t\right)\phi_n(x),$$

or linear combinations of such solutions.

But what is the meaning of the constant E_n? Let the operator $(-\hbar/i)(\partial/\partial t)$ act on a wave function to see what average value corresponds to this operator. It will turn out to be E_n,

$$\int dx\,\psi_n^*\left(\frac{-\hbar}{i}\frac{\partial}{\partial t}\right)\psi_n = \int dx\,\psi_n^* E_n \psi_n = E_n,$$

because

$$\frac{-\hbar}{i}\frac{\partial}{\partial t}\psi_n(x,t) = \phi_n(x)\frac{-\hbar}{i}\frac{\partial f(t)}{\partial t} = E_n f(t)\phi_n = E_n\psi_n(x,t)$$

and

$$\int \psi_n^*\psi_n\,dx = 1$$

(the wave function is normalized). On page 124, $(-\hbar/i)(\partial/\partial t)$ is identified as the ENERGY OPERATOR, so E_n is the energy associated with a system in the pure state whose wave function is $\phi_n(x)$. Often the expansion of wave functions in terms of the energy eigenfunctions $\phi_n(x)$ is particularly useful.

Dirac notation

This is a good place to give the standard Dirac notation for operators, vectors, and dot products. The state of a system is given by a vector in Hilbert space denoted by $|\psi(t)\rangle$, which corresponds to the usual wave function ψ in the spatial-coordinate representation. This vector can be multiplied by a scalar without altering the state of the system; thus $|\psi(t)\rangle$ and $c|\psi(t)\rangle$ represent the same state. A wave function is also sometimes written as $|x\rangle$ rather than $|\psi\rangle$ and can be represented with weights w_n in terms of basis vectors $|u_n\rangle$ (the same as the eigenfunction u_n used just above) as

$$|x\rangle = \sum w_n |u_n\rangle,$$

or (for a continuous spectrum) as

$$|x\rangle = \int w_n |u_n\rangle \, dn,$$

while the eigenvector equation above would be

$$\mathbf{A}|u_n\rangle = a_n |u_n\rangle$$

($\mathbf{A}$ is an operator). The average value of property A (which corresponds to operator $\mathbf{A}$) is given by the inner (dot) product of $\langle x|$ and $\mathbf{A}|x\rangle$:

$$\langle a\rangle = \langle x|\mathbf{A}|x\rangle \equiv \int dx\, \psi^* \mathbf{A} \psi$$

Note that the complex conjugate of $|x\rangle$ is denoted by $\langle x|$. These relationships are summarized with *Dirac notation*:

DIRAC NOTATION

Standard	Dirac
$\psi(x,t)$	$|x\rangle$ or $|\psi\rangle$
$\psi^*(x,t)$	$\langle x|$ or $\langle\psi|$

$$u_n \qquad |u_n\rangle$$

$$w_n \qquad \langle\psi|u_n\rangle$$

$$\int \psi^* \mathbf{A}\psi \, dx \qquad \langle\psi|\mathbf{A}|\psi\rangle$$

Sometimes the wave function (a delta function) that places the particle definitely at a particular location x is denoted by $|x\rangle$, so the meaning of $|x\rangle$ can vary with the context. Some other relevant attribute, such as the index assigned to a state ($|13\rangle$, for example) might also appear in $|\ \rangle$.

Now is a time to take a step back and survey our position. Every measurable property a has been associated with an operator **A**, defined above. The average value of the property is $\langle a\rangle = \langle x|\mathbf{A}|x\rangle$. What classical physics measures is these average values, and the correspondence principle requires that the averages predicted by quantum mechanics should satisfy the laws of classical mechanics. If a classical law states that $a = b$, then the correspondence principle requires that in the quantum-mechanical formulation (with operators **A** and **B**), this same relation hold for the average values:

$$\langle x|\mathbf{A}|x\rangle = \langle x|\mathbf{B}|x\rangle$$

This means that the quantum-mechanical law

$$\mathbf{A}\psi = \mathbf{B}\psi$$

in turn corresponds to the classical law $a = b$. Thus the rule for going from classical mechanics to quantum mechanics is: Replace all classical variables in the classical equation by their operators (acting on the wave function). As we will see later, some of the operators do not commute, whereas the quantities in classical equations always commute. Thus, following this prescription will require care in some cases.

At this point we have found abstract expressions for operators that are associated with measurable quantities but have not identified the operators explicitly. The next job is to give an illustration of how the preceding operator formalism is used. The goal is to find explicit expressions for the operators that correspond to momentum, position, and energy.

To find the operator **P** associated with the momentum measurement, we must identify the transition amplitude $K_p(x, t)$ that relates momentum amplitude to ψ. To do this, recall that

$$\phi(p) = \int K_p(x)\psi(x)\,dx,$$

and also that

$$\phi(p) = \int dx\,\psi(x)\exp\left(-\frac{ipx}{\hbar}\right)$$

(page 114). Thus the transition amplitude for the momentum measurement can be identified as

$$K_p(x) = \exp\left(-\frac{ipx}{\hbar}\right).$$

The kernel of the associated operator **P** is

$$\begin{aligned} P(x,\xi) &= \int pK_p(x)K_p^*(\xi)\frac{dp}{2\pi\hbar} \\ &= \int p\exp\left(-\frac{i}{\hbar}px\right)\exp\left(\frac{i}{\hbar}p\xi\right)\frac{dp}{2\pi\hbar} \\ &= \int p\exp\left(-\frac{i(x-\xi)p}{\hbar}\right)\frac{dp}{2\pi\hbar} \\ &= -\frac{\hbar}{i}\delta'(x-\xi). \end{aligned}$$

To follow these last manipulations, note two things: The first is that the probability of finding a momentum close to p is $|\phi|^2(dp/2\pi\hbar)$, so $2\pi\hbar$ must appear in the volume element (see the discussion of momentum measurement on page 113). The second is that the last step in the preceding chain of equations depends on the observation that the Fourier transform of a delta function is one: $\mathscr{F}\{\delta(x)\} = 1$. This means that

$$\mathscr{F}^{-1}\{1\} = \int \frac{1}{2\pi}e^{i\omega x}\,d\omega = \delta(x)$$

and

$$\frac{d}{dx}\mathscr{F}^{-1}\{1\} = \int \frac{i\omega}{2\pi}e^{i\omega x}\,d\omega = \frac{d\delta(x)}{dx} \equiv \delta'(x).$$

Finally, identify $\omega = p/\hbar$ to give

$$\mathbf{P}\psi(x) = \int d\xi \frac{-\hbar}{i}\delta'(x-\xi)\psi(\xi) = \underbrace{\frac{\hbar}{i}\partial_x}_{\mathbf{P}}\psi(x)$$

(use integration by parts for the last step). So the momentum operator is

$$\mathbf{P} = \frac{\hbar}{i}\frac{\partial}{\partial x}.$$

We now have an explicit expression for the momentum operator. How about the position and energy operators? Recall that for operator $\mathbf{A}$, the average value of the corresponding property is $\langle x|\mathbf{A}|x\rangle$; this is just the way that the operator was defined earlier. One way to identify an operator, then, is to carry out a procedure like the one just done for momentum. An alternative is to use the definition $\langle a\rangle = \langle x|\mathbf{A}|x\rangle$ and identify the operators that satisfy this relationship. This approach leads one at once to the identities of the position operator and of the potential-energy operator. By definition, the average position of a particle (for probability $P(x)$) is

$$\langle x\rangle = \int dx\, P(x)x = \int dx\, |\psi(x)|^2 x = \int dx\, \psi^* x\psi \equiv \langle x|\mathbf{X}|x\rangle.$$

So the position operator is just

$$\mathbf{X} = x.$$

Similarly, the average potential energy of a particle is

$$\langle V\rangle = \int dx\, P(x)V(x) = \int dx\, |\psi(x)|^2 V(x) = \int dx\, \psi^* V(x)\psi \equiv \langle x|\mathbf{V}|x\rangle,$$

and the potential energy operator is

$$\mathbf{V} = V.$$

In three dimensions, the key relations above are

$$\phi(\mathbf{p}) = \int e^{-i\mathbf{p}\cdot\mathbf{x}}\psi(\mathbf{x})\, d^3x,$$

$$\psi(\mathbf{x}) = \int e^{i\mathbf{p}\cdot\mathbf{x}}\frac{d^3p}{(2\pi\hbar)^3},$$

$$\mathbf{P} = \frac{\hbar}{i}\nabla\cdot,$$

$$\mathbf{X} = \mathbf{x},$$

$$\mathbf{V} = V(\mathbf{x}).$$

Finally, the energy operator. The operator that corresponds to momentum is a spatial derivative. But the wave function depends jointly on space and on time. What corresponds to the time derivative of ψ? We will see that the partial derivative with respect to time is the energy operator (up to a proportionality constant). The way to find the total energy operator is to start with the classical equation for the energy: $E = p^2/2m + V$ (for a particle with mass m moving in a potential V). Following the prescription that operators are to replace observable quantities, we have that

$$p^2/2m + V \Rightarrow -\frac{\hbar^2}{2m}\frac{\partial^2}{\partial x^2} + V = \mathbf{H}.$$

This operator version of the equation for the total energy can at once be identified as the Hamiltonian (defined on page 112) that appears in SCHRÖDINGER'S EQUATION:

$$-\frac{\hbar}{i}\frac{\partial\psi}{\partial t} = \mathbf{H}\psi = -\left(\frac{\hbar}{2m}\frac{\partial^2}{\partial x^2} - V\right)\psi$$

Thus we can identify the operator **E** that corresponds to the particle's energy as

$$\mathbf{E} = \frac{-\hbar}{i}\frac{\partial}{\partial t}.$$

In summary, for a particle in one spatial dimension, the *momentum operator*, *position operator*, and *energy operator* are as follows:

MOMENTUM OPERATOR

$$\mathbf{P} = \frac{\hbar}{i}\frac{\partial}{\partial x}$$

POSITION OPERATOR

$$\mathbf{X} = x$$

ENERGY OPERATOR

$$\mathbf{E} = -\frac{\hbar}{i}\frac{\partial}{\partial t}$$

The Uncertainty Principle

In an earlier section we saw that a particle's motion can be described in either the position or momentum representations and that these two representations are related by Fourier transforms. This fact has an interesting consequence: because of a property of Fourier transforms, the more concentrated the position amplitude $\psi(x)$ is around a particular spatial location, the more spread out the momentum amplitude $\phi(p)$. In the limiting case, if $\psi(x) = \delta(x)$, then $\phi(p) = 1$; that is, the momentum is equal everywhere in momentum space. The converse is also true: if the position amplitude is very spread out, the momentum amplitude is concentrated around one point in momentum space. Since the probability of finding the particle at a particular location is given by $|\psi(x)|^2$, knowing a particle's likely spatial location means that we have only a very imprecise idea of its momentum. This notion of a reciprocal relationship between knowledge of position and knowledge of momentum gives rise, when it is made more precise, to the famous *Heisenberg uncertainty principle*:

HEISENBERG
UNCERTAINTY PRINCIPLE

$$\Delta x \Delta p = \tfrac{1}{2}\hbar$$

Here Δx is the uncertainty (standard deviation) in the particles's position, and Δp the uncertainty in its momentum. This uncertainty principle will hold, it will turn out, for any pair of *canonically conjugate* variables. In classical mechanics, this concept is defined with Poisson brackets, so any pair of variables for which $\{a, b\} = 1$ are canonically conjugate (see page 68). The analogous relation in quantum mechanics, for a pair of operators **A** and **B**, uses the *commutator*:

COMMUTATOR

$$[\mathbf{A}, \mathbf{B}] \equiv \mathbf{AB} - \mathbf{BA}$$

So **A** and **B** are canonically conjugate if $[\mathbf{A}, \mathbf{B}] = i\hbar$, and otherwise **A** and **B** commute, as $[\mathbf{A}, \mathbf{B}] = 0$.

The goal of this section is to see how the uncertainty principle and the concept of the commutator arise. The reader who is not interested in the details of how the uncertainty principle arises can skip from this point to the next section.

Call the position and momentum variances Δx^2 and Δp^2. So by definition,

$$\Delta x^2 \equiv \int dx\, \psi^* |\mathbf{X} - \langle \mathbf{X} \rangle|^2 \psi = \int dx\, |(\mathbf{X} - \langle \mathbf{X} \rangle)\psi|^2$$

and

$$\Delta p^2 \equiv \int dx\, \psi^* |\mathbf{P} - \langle \mathbf{P} \rangle|^2 \psi = \int dx\, |(\mathbf{P} - \langle \mathbf{P} \rangle)\psi|^2.$$

The first of these equation chains used the fact that $\mathbf{X} = x$ is real, and the second used integration by parts. The product, then, is

$$\Delta x^2 \Delta p^2 \equiv \int dx\, |(\mathbf{X} - \langle \mathbf{X} \rangle)\psi|^2 \int dx\, |(\mathbf{P} - \langle \mathbf{P} \rangle)\psi|^2.$$

The Schwartz inequality is

$$\int dx\, |f|^2 \int dx\, |g|^2 \geq \left| \int dx\, f^* g \right|^2.$$

This can be used to rewrite the $\Delta x^2 \Delta p^2$ equation as the inequality

$$\Delta x^2 \Delta p^2 \geq \left| \int dx\, [(\mathbf{X} - \langle \mathbf{X} \rangle)\psi]^* (\mathbf{P} - \langle \mathbf{P} \rangle)\psi \right|^2.$$

Because $\mathbf{X} = x$, this last equation becomes

$$\Delta x^2 \Delta p^2 \geq \underbrace{\left| \int dx\, \psi^* (\overline{\mathbf{X}}\overline{\mathbf{P}}) \psi \right|^2}_{(1)},$$

where

$$\overline{\mathbf{X}} = \mathbf{X} - \langle \mathbf{X} \rangle \quad \text{and} \quad \overline{\mathbf{P}} = \mathbf{P} - \langle \mathbf{P} \rangle.$$

The next step is to expand the integral in (1) by writing the operators in its integrand as

$$\bar{\mathbf{X}}\bar{\mathbf{P}} = \tfrac{1}{2}(\bar{\mathbf{X}}\bar{\mathbf{P}} - \bar{\mathbf{P}}\bar{\mathbf{X}}) + \tfrac{1}{2}(\bar{\mathbf{X}}\bar{\mathbf{P}} + \bar{\mathbf{P}}\bar{\mathbf{X}}).$$

So the integral in (1) expands to three terms, just as $|a + b|^2 = |a|^2 + |b|^2 + |ab + ba|$. The cross term cancels because

$$\underbrace{\left|\int dx\, \psi^* \bar{\mathbf{X}}\bar{\mathbf{P}}\psi\right|^2}_{(1)} = \left[\int dx\, \psi^* \bar{\mathbf{X}}\bar{\mathbf{P}}\psi\right]\left[\int dx\, \psi^* \bar{\mathbf{X}}\bar{\mathbf{P}}\psi\right]^*$$
$$= \left[\int dx\, \psi^* \bar{\mathbf{X}}\bar{\mathbf{P}}\psi\right]\left[\int dx\, \psi \bar{\mathbf{X}}^* \bar{\mathbf{P}}^* \psi^*\right]$$
$$= \left[\int dx\, \psi^* \bar{\mathbf{X}}\bar{\mathbf{P}}\psi\right]\left[\int dx\, \psi \bar{\mathbf{X}}\bar{\mathbf{P}}^* \psi^*\right]$$
$$= \left[\int dx\, \psi^* \bar{\mathbf{X}}\bar{\mathbf{P}}\psi\right]\left[\int dx\, \psi^* \bar{\mathbf{P}}\bar{\mathbf{X}}\psi\right].$$

The manipulations here depended on the fact that $\bar{\mathbf{X}} = x - \langle x \rangle$ is real and, for the last step where integration by parts was used, that $\bar{\mathbf{P}}^* = -\bar{\mathbf{P}}$. The first square term in this expansion of (1) is

$$\frac{1}{4}\left|\int dx\, \psi^*(\bar{\mathbf{X}}\bar{\mathbf{P}} + \bar{\mathbf{P}}\bar{\mathbf{X}})\psi\right|^2,$$

and this is a real number ≥ 0. Thus

$$\Delta x^2 \Delta p^2 \geq \frac{1}{4}\left|\int dx\, \psi^*(\bar{\mathbf{X}}\bar{\mathbf{P}} + \bar{\mathbf{P}}\bar{\mathbf{X}})\psi\right|^2,$$

where the integral on the right is the third (square) term in the expansion of (1). A relation of the form for the operators that appears in the integrand provides the definition of a COMMUTATOR:

$$[\mathbf{A}, \mathbf{B}] \equiv \mathbf{AB} - \mathbf{BA}$$

Since we know the operators in this case, we can evaluate it explicitly:

$$[\mathbf{X}, \mathbf{P}] = \frac{\hbar}{i}\left(x\frac{\partial}{\partial x} - \frac{\partial}{\partial x}x\right) = i\hbar$$

Here the average values cancel, so they do not appear, and the last step depended on integration by parts. The product of the variances is

$$\Delta x^2 \Delta p^2 = \tfrac{1}{4}\hbar^2.$$

Finally, taking the square root of this, we have the HEISENBERG UNCERTAINTY PRINCIPLE:

$$\Delta x \Delta p = \tfrac{1}{2}\hbar$$

The commutator [**X**, **P**] does not vanish, because these operators do not commute: measuring position first and then momentum in general gives a different answer than the same measurements made in the reverse order because determining position necessarily, by virtue of the uncertainty principle, makes momentum indeterminate. This lack of commutativity is not true for all operators. For example, as is quickly verified, the commutator of the operators for x and y positions $[\mathbf{X}, \mathbf{Y}] = 0$; these operators commute.

In chapter 2, on classical mechanics, the POISSON BRACKET was seen to define canonically conjugate variables (page 68). The commutator is the quantum-mechanical version of the Poisson bracket, since it follows the same algebra and, as seen for momentum and position variables, identifies the corresponding operators as conjugate. Specifically, a pair of operators $\mathbf{A}_j$ and $\mathbf{B}_j$ that correspond to canonically conjugate variables (a_j, b_j) do not commute, and the Poisson bracket $\{a_j, b_k\} = \delta_{jk}$ is replaced by the commutation relation

$$[\mathbf{A}_j, \mathbf{B}_k] = i\hbar\delta_{jk}.$$

In the next section we will see that the commutator plays a central role in the Heisenberg picture of quantum mechanics, an alternative to Schrödinger's equation that provides a basis for quantizing systems where the guidance of classical mechanics is unavailable.

The Schrödinger and Heisenberg Pictures

Although, as noted above, the Schrödinger equation is very convenient for many practical problems, it does not form a suitable basis for extending quantum mechanics to other situations. Consider, for example, an attempt to develop a quantum-mechanical theory for the interaction of charged particles with the electromagnetic field. An approximate treatment could use quantum mechanics for the particles and classical theory for the field. But a complete theory would have to deal explicitly with the fact that the electromagnetic field is itself quantized, as is shown by the photoelectric effect, for example. Because the Lagrangian of the electromagnetic field can be found (see the subsection "The covariant form of electricity and magnetism," page 194), the Feynman approach, which expresses the quantum behavior of a system in terms of its Lagrangian, provides a natural way to develop a quantum description of this field. An alternative formulation of quantum mechanics that provides a route for generalization to nonmechanical systems is the Heisenberg picture. The goal of this section is to introduce this version of quantum mechanics. The starting point is the evolution operator derived from Schrödinger's equation.

SCHRÖDINGER'S EQUATION can be solved formally for ψ to give the quantum mechanical *evolution equation*:

EVOLUTION EQUATION
(QUANTUM MECHANICS)

$$\psi(t) = \exp\left(-\frac{i}{\hbar}t\mathbf{H}\right)\psi(0)$$

This equation thus relates the wave function at an initial time to that at later times and identifies the *evolution operator*:

EVOLUTION OPERATOR

$$\exp\left(-\frac{i}{\hbar}t\mathbf{H}\right)$$

(See the discussion on page 14 for the definition of functions of operators.)

In the version of quantum mechanics presented so far, the operators have been time invariant, and the state of the system, described by the amplitude $\psi(t)$, has depended on time. An alternative representation, the *Heisenberg picture*, places the time dependence in the operators. To see how this time dependence of operators can arise, consider a general operator $\mathbf{A}_0$ (the momentum operator, for example) that is, of course, time-independent. The average value of the quantity that corresponds to $\mathbf{A}_0$ is given by

$$\langle\psi(t)|\,\mathbf{A}_0\,|\psi(t)\rangle = \underbrace{\langle\psi(0)|\exp\left(\frac{i}{\hbar}t\mathbf{H}\right)}_{\langle\psi(t)|}\mathbf{A}_0\underbrace{\exp\left(\frac{i}{\hbar}t\mathbf{H}\right)|\psi(0)\rangle}_{|\psi(t)\rangle},$$

because, according to the evolution equation, $|\psi(t)\rangle = \exp(-(i/\hbar)t\mathbf{H})|\psi(0)\rangle$, and its complex conjugate is $\langle\psi(t)| = \langle\psi(0)|\exp((i/\hbar)t\mathbf{H})$. Now define the operator $\mathbf{A}(t)$ as follows:

$$\mathbf{A}(t) = \exp\left(\frac{i}{\hbar}t\mathbf{H}\right)\mathbf{A}_0\exp\left(-\frac{i}{\hbar}t\mathbf{H}\right)$$

This operator is time-dependent, and it acts on a time-independent amplitude $\psi(0)$ that specifies a state of the system. When the time-dependent operators are used, we have the Heisenberg view, and when the time-dependence is in the amplitudes, the Schrödinger picture is being used.

Both views are, of course, physically equivalent. The utility of the Heisenberg picture arises from the fact that a differential equation for operators replaces Schrödinger's equation. Consider a small time displacement from t to $t + \tau$ (τ is small). The operator $\mathbf{A}(t)$ thus satisfies, to first order, the following equation:

$$\begin{aligned}\mathbf{A}(t+\tau) &= \mathbf{A}(t) + \tau\frac{d\mathbf{A}(t)}{dt}\\ &= \exp\left(\frac{i}{\hbar}\tau\mathbf{H}\right)\mathbf{A}(t)\exp\left(-\frac{i}{\hbar}\tau\mathbf{H}\right)\\ &= \left(1 + \frac{i}{\hbar}\tau\mathbf{H}\right)\mathbf{A}(t)\left(1 - \frac{i}{\hbar}\tau\mathbf{H}\right)\\ &= \mathbf{A}(t) + \frac{i}{\hbar}\tau\mathbf{H}\mathbf{A}(t) - \frac{i}{\hbar}\tau\mathbf{A}(t)\mathbf{H}\end{aligned}$$

So

$$\frac{d\mathbf{A}(t)}{dt} = \frac{i}{\hbar}(\mathbf{HA} - \mathbf{AH}).$$

By recognizing that $(\mathbf{HA} - \mathbf{AH}) = [\mathbf{H}, \mathbf{A}]$, we get the *operator equation of motion*:

OPERATOR EQUATION OF MOTION

$$\frac{d\mathbf{A}(t)}{dt} = \frac{i}{\hbar}[\mathbf{H}, \mathbf{A}]$$

Note that this equation is more like the usual classical-mechanics equations of motion: operators correspond to dynamic variables, and Hamilton's equations describe the motion of these variables. In fact, the operator equation of motion is just the quantum mechanical version of Hamilton's equations expressed with Poisson brackets (see the section "Poisson Brackets," page 68).

To quantize a novel system, then, one could describe the system by a Lagrangian and use the Feynman formulation. Alternatively, commutation relations can be developed, and the operator equation of motion used to describe the kinetics of the system. Just how this program is carried out is the subject of field theory.

Time-Varying Forcing*

We now turn to the issue of how a particle behaves under an external driving force. Picture, for example, a driven harmonic oscillator. This question is important for a variety of reasons. First, the preceding treatment has used initial conditions to determine the path a system follows, but sometimes one wishes to start all systems in a standard state and impose a response through the action of some driving function. The treatment that follows gives a way of eliminating the need for establishing particular initial conditions case by case. Second, the number of functional integrals that can be explicitly solved is very small, and practical problems generally require the use of approximations. This question of forcing functions provides the appropriate context for a discussion of such approximations. Finally, the calculations to be presented are just the ones that will be used for the development of quantum field theory in chapter 7.

This material is best presented in the context of a specific example, the driven harmonic oscillator mentioned above. The classical Lagrangian is the kinetic ($m\dot{x}^2/2$) minus the potential energy $V(x,t)$:

$$L = \frac{m}{2}\dot{x}^2 - V(x,t)$$

For notational convenience, take the mass $m = 1$. For a free harmonic oscillator, $V(x, t) = \omega^2 x^2$, and for a driven oscillator,

$$V(x, t) = \tfrac{1}{2}\omega^2 x^2 - J(t)x(t)$$

for a time-varying forcing function $J(t)$.

To see why J works as a forcing function here, consider a general Lagrangian

$$L_0 + J(t)x(t),$$

where L_0 is the Lagrangian of the undriven system and J is the driving function. Look at the effect of J alone. The value of x for which the action is an extremum dominates the functional integral that determines the transition amplitudes. If the action were only

$$\int J(t)x(t)\,dt,$$

then the extremum would be achieved when x is given by the functional derivative

$$\frac{\delta}{x(t)} \int J(t)x(t)\,dt = J(t)$$

(see page 34). Thus x tends to have the value equal to J; that is, J forces x.

We wish to start the oscillator in its lowest-energy state (the *vacuum* state) at $t = -\infty$, have it finish in the same state at $t = \infty$, and observe the effect of driving by a specified function $J(t)$ at some times near 0 (the time we do the experiment). This transition amplitude (vacuum state to vacuum state, with driving in between) is denoted by $W[J]$ and is given by

$$W[J] = \int \mathscr{D}x(t) \exp\left(\frac{i}{\hbar} \int [L_0 + J(t)x(t)]\,dt\right) = \int \mathscr{D}x \exp\left(\frac{i}{\hbar} S\right),$$

where S is the action and L_0 is, in this specific example, the free harmonic-oscillator Lagrangian:

$$L_0 = \frac{m}{2}(\dot{x}^2 - \omega^2 x^2)$$

$W[J]$ is known as a *generating functional.*

The goal of our manipulations will be to express $W[J]$ in terms of $W[0]$ (the generating functional for the undriven system) and a functional (to be discovered) that involves J but not $x(t)$. Note that in this specific example, $W[0]$ can be calculated exactly because it is just a Gaussian integral. The representation we seek thus forms the basis for finding an approximate answer for the integral $W[J]$.

Our manipulations will require that we express L_0 in terms of $x(t)$; that is, we need to find an operator, call it $\mathbf{M}$, that operates on x to give us $-L_0$. The required operator comes from the Euler-Lagrange equation that corresponds to L_0. To find this equation, apply $\partial_x - (d/dt)\partial_{\dot{x}}$ to L_0 (see page 61). This operation gives the Euler-Lagrange equation (specifically, the wave equation) for the harmonic-oscillator Lagrangian:

$$\left(\partial_x - \frac{d}{dt}\partial_{\dot{x}}\right)\underbrace{\tfrac{1}{2}(\dot{x}^2 - \omega^2 x^2)}_{L_0} = -\left(\frac{d^2x(t)}{dt^2} + \omega^2 x(t)\right)$$

$$\underbrace{\left(\frac{d^2}{dt^2} + \omega^2\right)}_{\mathbf{M}} x(t) \equiv \mathbf{M}x(t) = 0$$

Using integration by parts and the explicit expression for $\mathbf{M}$, one can verify that

$$-\frac{1}{2}\int dt\, x(t)\mathbf{M}x(t) = \int dt\, L_0(x, \dot{x}),$$

which gives the desired expression for L_0 in terms of x. Specifically, for the harmonic oscillator,

$$\begin{aligned}
-\int dt\, x(t)\mathbf{M}x(t) &= -\int dt\, x(t)\left(\frac{d^2}{dt^2} + \omega^2\right)x(t) \\
&= -\int dt\, x(t)\left(\frac{d^2x}{dt^2} + \omega^2 x^2(t)\right) \\
&= -\int dt\left(x(t)\frac{d^2x}{dt^2} + \omega^2 x^2(t)\right)
\end{aligned}$$

$$= -\int dt \left(-\left(\frac{dx}{dt}\right)^2 + \omega^2 x^2(t) \right)$$

$$= 2 \int dt\, L_0(x, \dot{x}).$$

We have used a particular Lagrangian to obtain this relationship, but it holds for a larger class of Lagrangians. The basis for this is the fact that the Euler Lagrange equation is given by the functional differential equation $\delta S/\delta y = 0$ (see page 38). If $S = \frac{1}{2}\int dt\, y\mathbf{M}y$, then $\delta S/\delta y = \mathbf{M}y$ for a variety of operators $\mathbf{M}$.

The operator in the Euler-Lagrange equation has an inverse, the impulse response (Green's function) $-D(t)$, that obeys the equation

$$\mathbf{M}D(t) = \delta(t).$$

The trick we use, which will separate out the influence of J, is to make a change of variables in the integral $W(J)$:

$$x(t) \to x'(t) = x(t) - \int d\tau\, D(t-\tau)J(\tau)$$

So with the shorthand $D * J$ for the convolution,

$$x(t) = x'(t) + D * J.$$

Note that $D(t)$ here is the Green's function for the Euler-Lagrange-equation operator $\mathbf{M}$. The change of variable amounts to the addition of a constant in function space, so $\mathscr{D}x = \mathscr{D}x'$. With this change of variable (and the notation $D * J$ for the convolution in the preceding equation that defines the change of variable), the action becomes

$$S = -\frac{1}{2}\int dt\,[x'(t) + D * J]\mathbf{M}[x'(t) + D * J] + \int dt\,[x'(t) + D * J]J$$

$$= \underbrace{-\frac{1}{2}\int dt\, x'\mathbf{M}x'}_{(1)} \underbrace{-\frac{1}{2}\int dt\, x'\mathbf{M}D * J}_{(2)} \underbrace{-\frac{1}{2}\int dt\, D * J\mathbf{M}x'}_{(3)}$$

$$\underbrace{-\frac{1}{2}\int dt\,[D * J][\mathbf{M}D * J]}_{(4)} + \underbrace{\int dt\, Jx'}_{(5)} + \underbrace{\int dt\, JD * J}_{(6)}.$$

Let us look at the integrals one by one. Integral (1) is just the action of an undriven oscillator and is replaced by the integral:

$$(1) = \frac{1}{2}\int dt\, L_0(x', \dot{x})$$

For integral (2), write out the convolution (remembering that $\mathbf{M}D(t) = \delta(t)$):

$$(2) = \frac{1}{2}\int dt\, x'(t)\int d\tau\, \mathbf{M}D(t-\tau)J(\tau)$$

$$= \frac{1}{2}\int dt\, x'(t)\int d\tau\, \delta(t-\tau)J(\tau)$$

$$= \frac{1}{2}\int dt\, J(t)x'(t)$$

Thus integral (2) cancels half of integral (5). An analogous calculation for integral (3) (except that integration by parts is needed to move $\mathbf{M}$ from x' to D) shows that (3) cancels the other half of integral (5). Thus we are left with integrals (4) and (6). The $\mathbf{M}D$ in integral (4) gives a δ function, so

$$(4) = \frac{1}{2}\int dt\, JD * J,$$

and this cancels half of integral (6). The final expression for the action, then, is

$$S = \int dt\, L_0(x', \dot{x}) + \frac{1}{2}\int dt\, JD * J.$$

The last integral here is a constant with respect to the functional integral because it depends only on J and D, and not at all on the integration variable x, so it comes outside the functional integral. Thus $W[J]$ is the *generating functional*:

GENERATING FUNCTIONAL

$$W[J] = W[0]\exp\left(\frac{i}{2\hbar}\int\int dt d\tau\, J(t)J(\tau)D(t-\tau)\right)$$

Here

$$W[0] = \int \mathscr{D}x(t) \exp\left(\frac{i}{\hbar} \int dt\, L_0(x, \dot{x})\right).$$

The $W[J]$ that describes the driven system is therefore expressed in terms of the behavior of the unperturbed oscillator (a Gaussian integral that we can evaluate) and another functional that involves only the Green's function for the Euler-Lagrange operator and the driving function J.

5 Statistical Physics

Some areas of physics, classical mechanics or special relativity, for example, are really elaborations of just one or two ideas, and the theory is naturally presented as a direct development of these ideas. Other parts of physical theory have been succinctly codified by an individual, like Feynman's version of quantum mechanics, and a description of such a theory from this particular point of view is coherent and unified. Statistical physics, however, is perhaps best viewed in its historical context because thermodynamics does not readily lend itself to an intuitive understanding and statistical mechanics differs from the rest of classical physics because it provides relations between stochastic characteristics of aggregates rather than laws that apply to individual entities.

Thermodynamics developed in the middle of the nineteenth century and was largely the product of two intellectual trends: the first was an attempt by nineteenth-century physicists to unify all branches of physics, and the second was a drive to develop conservation laws.

Although the historical statements in this presentation are reasonably accurate, what follows resembles more an historical novel than a history of thermodynamics; the history is presented to motivate the final formulation of statistical physics rather to illuminate the interesting evolution of ideas involved.

Historical Context

Where did physics stand at the middle of the nineteenth century? Classical mechanics, as described in chapter 2, was fully developed and recognized as such. The main experimental observations on which J. C. Maxwell (1831–1879) was to base his theory of electricity and magnetism had been made—the definitive version of Maxwell's equations was published in 1873—and many of the properties of light were known. Considerable effort was being devoted to the study of heat. Experimental methods were available for measuring temperature and the flow of heat (by temperature changes in a reservoir), although the physical basis of heat was still mysterious. The caloric theory, the view that heat was a weightless substance that flowed from high to low temperature regions, was no longer widely accepted, because the theory had unpalatable consequences. For example, caloric (a substance) could be created in unlimited quantities by friction and could cause ice to melt without changing its temperature. The caloric theory was not a success, and neither were its midcentury rivals: one held that heat was a wave, like light (think of being warmed by sunlight), and the other held that it was somehow related to the motions of hypothetical atoms (whose existence was doubted until the turn of the century). One theoretical treatment of heat in the modern sense was available, and this theory brought the study of heat into the mainstream of midcentury mathematical physics: Fourier had developed a differential equation (the heat equation) that describes the flow of heat through a body.

Two workers can be identified as providing the principal basis for the midcentury development of thermodynamics: J. P. Joule (1818–1889) and S. Carnot (1796–1832). The modern concept of work, and its relation to a system's energy, had been established, and Joule had shown experimentally that electrical and mechanical work could be converted into heat, and that a fixed quantity of heat was associated with a fixed quantity of work, whether electrical or mechanical. Joule's experiments were widely known and appreciated because they demonstrated that heat, mechanics, and electricity (light was already viewed as somehow related to electricity because both were thought to travel through the same ether) were in some way connected, so a grand unified theory should be possible. Naturally, such a theory had to be related to mechanics, because this was the only fully developed physical theory available at the time.

Carnot was dead by midcentury, but he had published (in 1824) an important and strikingly creative analysis of the efficiency of steam engines, which B. P. E. Clapeyron had reformulated (in 1834) in a more standard way. Carnot developed a description of an idealized engine that, after four steps (the Carnot cycle), was returned to its starting state, having converted heat flow into work. His idealized engine carried out its steps reversibly, that is, so slowly that any step could be run in the reverse direction to end up where it had started without using any energy. Further, Carnot demonstrated that no real engine could be more efficient than his idealized engine. His argument for the engine's efficiency was a quite modern one: Suppose that a real engine were more efficient than Carnot's idealized one. Then if a real engine were coupled to the idealized engine, the engine pair could transfer heat from a cold to hot body without requiring any work to accomplish this transfer, a situation contrary to experience. Therefore, an analysis of the ideal engine placed limits on the efficiency of real engines. Carnot accepted the caloric theory in his treatment, and he assumed that the quantity of caloric was conserved around each cycle. His idea was that a heat engine is like a water engine that converts the flow of water into work without using up any of the water; by analogy, Carnot thought that the caloric runs down hill (from high to low temperatures) to power the heat engine, but that none of the "fluid" was used up.

The drive for conservation laws can be traced to the impact of Newtonian physics on nineteenth-century thought. It seemed to the seventeenth- and eighteenth-century intellectual that mechanics could explain everything, and that, to use the standard cliché, God was just a master clock maker who created a world and let it run. But why did the clock not run down? A common answer was that motion was conserved. Leibniz, for example, had claimed that *vis viva* (kinetic energy, in modern terminology) could be converted into other motion, but the total quantity in the world was constant. As nineteenth-century physics became successful in accounting for aspects of the physical world beyond mechanics, the natural extension of this seventeenth- and eighteenth-century worldview was the unification of various areas of physical knowledge so that everything became part of mechanics. Conservation laws, particularly ones that connected the diverse parts of physical theory to the highly developed theory of mechanics, not only served to unify but also continued the intellectual tradition of identifying conserved quantities that kept the more complicated nineteenth-century world-clock from running down.

Thermodynamics

The first law

The midcentury scientist R. Clausius (1822–1888) can, for our purposes, be singled out as having formulated thermodynamics, although the contributions of others were, of course, required. Clausius recognized that Carnot's treatment of heat engines, in which the flow of heat produces work, and Joule's experimental observations were in direct conflict because Carnot had assumed that the caloric fluid was conserved, whereas Joule had shown empirically that work and heat are interconvertible. Clausius accepted Joule's conclusions and formulated them as the *first law* of thermodynamics, describing the conservation of energy:

FIRST LAW

$$dU = dq + dW$$

Here dU is the increment in energy of a system, dq is the heat that flows into the system, and dW is the work done on the system. Note that after a quantity of heat dq is transferred to the system or an amount of work dW is done on the system, they are changed into energy (dU) and no longer exist as distinct quantities. Thus the system contains not a quantity of heat Q or an amount of work W but rather an energy U. For this reason, dq and dW are called *inexact differentials*, in contrast to the exact differential dU, which can be integrated to find the amount of energy in the system.

The second law

Clausius noticed that Carnot's analysis, which had explicitly supposed that the caloric is conserved, did not actually require this assumption of heat conservation; rather, the same analysis held if it was forbidden to have heat flow from colder to hotter regions (without work being done to produce the wrong-way transfer). Clausius's argument will be presented in detail below, but it can be summarized briefly as follows: Clausius noted that, instead of caloric conservation, Carnot only required that $dq/T \geq 0$ (T is the temperature). Because the quantity defined by dq/T was so important—this inequality is what defines the direction of heat flow—Clausius gave it the special name *entropy*. Clausius thus formulated the *second law* of thermodynamics:

SECOND LAW

$$dS \equiv \frac{dq}{T} \geq 0$$

Here S is the symbol designating entropy.

A consideration of heat engines (described below) demonstrates that two sorts of processes need to be recognized. A *reversible* process is an idealized one in which all changes are carried out so slowly that no energy is lost in friction or turbulence and the system is always in equilibrium. For such a process, any energy put in can be gotten out again, and the system can be returned exactly to its initial state without any losses of energy. An *irreversible* process—all real processes are like this—occurs spontaneously or is one in which some of the energy put in to drive the system is lost to friction and turbulence, so the system cannot be caused to exactly retrace its steps from the starting state. For a reversible process, the equality indicated in the second law holds (entropy is unchanged), whereas the entropy always increases for real, irreversible processes.

We turn now to an examination of the Carnot cycle and the second law of thermodynamics derived from it by Clausius. Start with two heat reservoirs, a hot one with temperature T_h and a cold one with temperature T_c, and a cylinder that contains one mole of ideal gas, that is, a gas for which the pressure p, the volume V, and the temperature T are related by the ideal gas law

$$pV = RT,$$

where R is the gas constant, (a law that was well known at the time). This cylinder is fitted with a frictionless piston that permits the experimenter to change the gas volume, and a thermometer that reports the temperature of the gas in the cylinder. The Carnot cycle describes an ideal heat engine that, it will turn out, sets an upper limit on the efficiency of all heat engines. The operation of a Carnot engine takes place in four reversible steps, each of which we will consider in turn.

Before starting our examination of the Carnot engine, however, we need an expression for the work performed by the changing volume of a gas. The units of volume are length^3, and the units of pressure are force/length^2. Therefore, pressure $\times$ volume has the units (force/length^2) $\times$ (length^3) = force $\times$ length, which is the units of work. Thus if we consider the pressure $p(V)$ of a gas to be a function of its volume V, the work ΔW done on the gas by a change in volume from V_a to V_b is given by

$$\Delta W = -\int_{V_a}^{V_b} p(V)\,dV.$$

The minus sign appears here because ΔW is the work done on the gas by permitting (or causing) a volume change. The first law, for a gas where the only work done is by the expansion and compression of the gas, is thus

$$dU = dq - p\,dV.$$

We now are ready to consider the four steps of the Carnot cycle:

1 Adiabatic compression Start with the system in state A. In this state the cylinder has volume V_A, pressure $p_A = RT_c/V_A$, and temperature T_c, because the cylinder has been in contact with the colder heat bath for a sufficient time to reach its temperature. The first part of the cycle involves removing the gas from contact with the cold reservoir and compressing it very slowly so that no work is done by creating turbulence. The gas is compressed to a volume V_B adiabatically (which means that the cylinder is isolated from the outside world so that no heat can flow into or out of the cylinder) and reversibly until its temperature is increased to T_h, the temperature of the hot reservoir. Work is required to accomplish this first step of the cycle. We must calculate what volume V_B is required to reach temperature T_h.

To perform this calculation, we need an expression for the heat capacity (as a function of the gas's volume) for an ideal gas; the heat capacity C_V relates the temperature change of the gas to the amount of heat that flows into it at a specified volume V. By definition, $C_V \equiv (dq/dT)_V$. Because the first law for a gas holds that $dq = dU - p\,dV$, if a particular volume V is specified and fixed, $dq = dU$, and

$$C_V = \left(\frac{dq}{dT}\right)_V = \left(\frac{\partial U}{\partial T}\right)_V$$

is the heat capacity at that volume. This relation means that

$$dU = C_V\,dT.$$

The heat capacity at a constant volume C_V could in principle vary with the volume of the gas, and in fact does vary a little for real gases. For an ideal gas (which approximates real gases under restricted circumstances) the heat capacity was known experimentally to be constant. For real gases, this heat capacity generally varies slowly over limited volume ranges.

For an infinitesimal change in volume around some specified volume V, the change in energy for an adiabatic compression (or expansion) is

$$dU = -p\,dV = -\frac{RT}{V}dV,$$

since, for a mole of ideal gas, $p = RT/V$. Equate these last two displayed expressions for dU:

$$C_V\,dT = -RT\frac{dV}{V},$$

or

$$\alpha\frac{dT}{T} = -\frac{dV}{V},$$

for $\alpha \equiv C_V/R$. Integrate this last relationship between V_A and V_B (with corresponding temperatures T_c and T_h),

$$\alpha \int_{T_c}^{T_h} \frac{dT}{T} = -\int_{V_A}^{V_B} \frac{dV}{V},$$

to get the relation between volume and temperature for an adiabatic compression (or expansion):

$$\alpha \ln\left(\frac{T_h}{T_c}\right) = -\ln\left(\frac{V_B}{V_A}\right) = \ln\left(\frac{V_A}{V_B}\right),$$

or, equivalently,

$$\frac{V_A}{V_B} = \left(\frac{T_h}{T_c}\right)^\alpha.$$

Note that it is important that C_V be constant for an ideal gas, for otherwise α would vary with V and the integration performed above would have to use the functional form of α. As a result of the first step in the Carnot cycle, then, we have performed a certain amount of work on the gas to compress it from its starting volume V_A to volume $V_B = (T_c/T_h)^\alpha V_A$ with a temperature T_h. This first step in the cycle has ended in state B with volume V_B, pressure p_B, and temperature T_h. Now we are ready for the second step.

2 Isothermal expansion Starting from B (with a volume V_B, pressure p_B, and temperature T_h), place the cylinder in contact with the hotter heat reservoir (*temperature* T_h, like that of the cylinder), and permit the piston to move slowly so that the gas can expand reversibly and do work. Throughout this expansion, the cylinder is maintained at temperature T_h by the transfer of heat from the heat reservoir (isothermal expansion). Permit the gas in the cylinder to reach volume V_C, with an associated pressure

$$p_C = \frac{RT_h}{V_C}$$

given by the ideal gas law. The work W_{BC} done on the environment by this expansion is

$$W_{BC} = \int_{V_B}^{V_C} p(V)\,dV = \int_{V_B}^{V_C} \frac{RT_h}{V}\,dV = RT_h \ln\left(\frac{V_C}{V_B}\right).$$

The amount of heat Q_1 that flows from the reservoir into the cylinder is, according to Joule, just equal to the work done, so

$$Q_1 = RT_\mathrm{h} \ln\left(\frac{V_C}{V_B}\right).$$

This second step in the cycle terminates in state C with volume V_C, pressure p_C, and temperature T_h.

3 Adiabatic expansion For the third step of the Carnot cycle, remove the cylinder from the heat reservoir so that it is thermally isolated (no heat can enter or leave), and permit a further reversible expansion (which does work on the environment) to the volume V_D associated with temperature T_c. The calculation for this adiabatic expansion is just like the one for the adiabatic compression in step (1), so volume V_D is given by

$$V_D = \left(\frac{T_\mathrm{h}}{T_\mathrm{c}}\right)^{\alpha} V_C.$$

After the third step, the system is in state D volume V_D, pressure p_D (given by the ideal gas law), and temperature T_c.

4 Isothermal compression Finally, place the cylinder (now at temperature T_c) in contact with the colder heat reservoir (also at temperature T_c), and slowly (and isothermally) compress the gas back to the starting volume V_A. During this reversible compression, a quantity of heat Q_2 is transferred from the gas to the colder heat reservoir, and the work W_{DA} done on the system, which transfers heat to the reservoir, is calculated like that for the isothermal expansion in step (2):

$$W_{DA} = RT_\mathrm{c} \ln\left(\frac{V_A}{V_D}\right) = -Q_2$$

The minus sign here signifies that the heat was transferred to the reservoir from the cylinder. At the end of this fourth step, the cylinder ends in state A, where it started.

The result of these four steps, then, has been to return the cylinder to its starting condition with volume V_A, pressure $p_A = RT_\mathrm{c}/V_A$, and temperature T_c. In the process, the quantity of work

$$W = W_{BC} - W_{DA} = Q_1 - Q_2$$

has been performed. The remainder of the analysis takes place in two steps. The first is to calculate the entropy change during the cycle (it will turn out to be 0), and the second is to show that any real cycle would have a larger entropy change. The key to this second argument will be the assertion that heat can never flow from a colder to a hotter body without work being done; thus the second law is what establishes the direction in which heat must flow. The first law says nothing about the direction of heat flow, and without the second law, heat could flow in either direction (hot to cold or cold to hot).

Because an entropy change for a reversible process ΔS is defined to be

$$\Delta S \equiv \frac{\Delta Q}{T}$$

for a process (where T is the temperature and ΔQ is the heat that flows during the process), entropy must remain constant for a reversible adiabatic process (in which $\Delta Q = 0$). Thus no entropy change occurred during the adiabatic steps (1) and (3) of the Carnot cycle. Entropy could change only during steps (2) and (4), the isothermal expansion and compressions. For the isothermal expansion, step (2), the entropy change ΔS_1 was

$$\Delta S_1 = \frac{Q_1}{T_h} = \frac{RT_h \ln(V_B/V_C)}{T_h} = R\ln\left(\frac{V_B}{V_C}\right),$$

and during isothermal compression, step (4), the entropy change ΔS_2 was

$$\Delta S_2 = \frac{-RT_c \ln(V_A/V_D)}{T_c} = -R\ln\left(\frac{V_A}{V_D}\right).$$

So the overall entropy change ΔS is

$$\Delta S = \Delta S_1 + \Delta S_2 = R\left(\ln\left(\frac{V_B}{V_C}\right) - \ln\left(\frac{V_A}{V_D}\right)\right).$$

But the volume changes associated with the adiabatic compression, step (1), are related by

$$\left(\frac{T_h}{T_c}\right)^{\alpha} = \frac{V_A}{V_B},$$

and the adiabatic expansion, step (4), by

$$\left(\frac{T_c}{T_h}\right)^\alpha = \frac{V_C}{V_D}.$$

Equate them, and you get

$$\frac{V_A}{V_B} = \frac{V_D}{V_C}.$$

This last equation can be rearranged to

$$\frac{V_A}{V_D} = \frac{V_B}{V_C},$$

which means that

$$-\ln\left(\frac{V_A}{V_D}\right) = -\ln\left(\frac{V_B}{V_C}\right).$$

Thus the entropy terms cancel, $\Delta S_1 = -\Delta S_2$, and entropy was unchanged by a turn of the Carnot cycle. Note, however, that if any heat had been wasted, that is, not turned into work, then the entropy change for the cycle $\Delta S = \Delta S_1 + \Delta S_2 > 0$. This is true because, for example, the heat Q taken from the reservoir in step (2) would be greater than the ideal Q_1, so the entropy Q/T_h would be greater than the ideal Q_1/T_h. The next step is to show that this is in general true for any real process.

We start by defining the efficiency ε of the Carnot engine:

$$\varepsilon \equiv \frac{W}{Q_1} = \frac{Q_1 - Q_2}{Q_1}$$

This efficiency measures the amount of work that we get out of the heat Q_1 absorbed by the engine (the cylinder) from the hotter reservoir. If all of the absorbed heat were converted into work, so that none was left to return to the colder reservoir, the efficiency would be 1. To demonstrate that any real process is actually less efficient that the Carnot engine, suppose the contrary: imagine that we have an engine that has a higher efficiency, so that it returns $Q_0 < Q_2$ to the colder reservoir. We could then couple this more efficient engine to the Carnot engine—which we could run backwards, because each step is carried out reversibly—and use its work to transfer $Q_0 - Q_2$ heat from the cold reservoir to the hot reservoir for each Carnot cycle. The result would then be to transfer heat from a colder to a hotter body without any work, and this is, of course, contrary to experience. Thus any real process must be no more efficient than the Carnot cycle. Generally, more heat must be used in some part of the process than would have been used in a Carnot cycle, so the entropy can only remain the same or increase. Note that I have framed the argument in terms of engines, as is traditional, but any physical process can be thought of as a part of a cycle, the remainder of the cycle being what is required to reversibly return the system to its starting state. Thus entropy must never decrease for any physical process.

The reason that entropy was identified as "something" is that it measures a defined quantity with an important property: it can only increase in real processes. The concept is a difficult one because of the way it arises as a ratio with a certain property (it never gets smaller) rather than a familiar characteristic of nature, like heat or work, with which we can make intuitive connections.

Work and energy were defined in the mid nineteenth century as they are now, but how are we to interpret heat and entropy? Because mechanical motion that does work on a system could be converted into heat, as Joule showed, Clausius assumed that heat was molecular motion, but he held that the laws of thermodynamics are independent of the precise mechanical interpretation of heat. Entropy was simply a defined quantity without an interpretation in terms of more elementary mechanical processes, although Clausius later had the idea that entropy was related in some way to the molecular configuration of the system (he was also one of the originators of the kinetic theory of gases).

Generalizing thermodynamics to more complex systems

The argument so far has been almost entirely in terms of heat engines, but, of course, thermodynamics applies to a wide range of systems in which two things are happening (or have the potential to happen): first, the system undergoes some change or transformation, and second, the flow of heat is involved. To see how this generalization starts, take another look at the first law: $dU = dq + dW$. According to the definition of entropy for a reversible process, $dq = T\,dS$, and this can be substituted into the first law applied to a gas ($dU = dq - p\,dV$) to give

$$dU(S, V) = T\,dS - p\,dV.$$

Note that according to this equation, the system's internal energy U must be a function of entropy and volume. In fact, by writing

$$dU(S,V) = \left(\frac{\partial U}{\partial S}\right)_V dS + \left(\frac{\partial U}{\partial V}\right)_S dV,$$

we can identify

$$T = \left(\frac{\partial U}{\partial S}\right)_V$$

and

$$p = -\left(\frac{\partial U}{\partial V}\right)_S.$$

Quantitites like U, S, and V, are called *extensive* variables because the value of the variable for a combination of two systems is the sum of the values of the variables for the individual systems. For example, if one system has a volume V_1 and another system has a volume V_2, then a third system made by combining these first two systems would have a volume $V_1 + V_2$. Variables like p and T are *intensive* because they have the same value throughout the system (if it is at equilibrium). The interesting point is that the differences in intensive variables between two subsystems act as driving forces for the change in the corresponding extensive variables. For example, a positive pressure difference between the inside of a balloon and the air outside will cause the volume of the balloon to increase.

These observations point the way for generalizing our formulation: for a system more complicated than the gas we have been considering, identify the additional extensive variables on which the internal energy depends, and then find the associated intensive variables. These new pairs contribute to the dW, term in the first law. For example, suppose that for the system we wish to study, we identify another extensive variable—call it X—with a corresponding intensive variable—call it f—given by $f = (\partial U/\partial X)_{S,V}$. The first law for this system would be

$$dU(S,V,X) = dq - p\,dV + f\,dX.$$

An example is provided by a system that consists of molecules of two different kinds: it contains N_1 moles of the first component and N_2 moles of the second. This example provides the basis for applying thermodynamics to chemical systems in which the evolution of the system involves chemical reactions that change the numbers of moles for various components. Define the *chemical potential* for the ith component as

$$\mu_i = \frac{\partial U}{\partial N_i},$$

where it is assumed that the other variables are held constant. The first law for this system now becomes

$$dU(S, V, N_1, N_2) = dq - p\,dV + \mu_1\,dN_1 + \mu_2\,dN_2.$$

What variables appear in the first law depend, of course, on the specific nature of the system and on what variables contribute to its internal energy. This ultimately must be determined by experiment, but the formalism will be the same in all cases.

Thermodynamic potentials

Knowledge of the energy U of a system allows one to calculate various quantities of interest. For example, for the simple ideal gas considered above, the pressure is found from $p = -\partial U/dV$. Because of the way experiments are done, variables known as *thermodynamic potentials* can be defined so that they exhibit many of the properties of the total internal energy U but are easier to deal with. One simplification in many experiments is that the system under study is, for reversible transformations, maintained at a constant temperature because it is in thermal contact with the fixed-temperature laboratory. Under this circumstance, the change in energy associated with a small step in some transformation that the system might undergo consists, as usual, of two components, one thermal and the other the work done on the system:

$$dU = T\,dS + dW$$

For the entire reversible evolution of the system, the total energy change (if, for convenience, we take the initial energy and entropy as 0) is just

$$U = \int_{\text{start}}^{\text{finish}} T(S)\,dS + W = TS + W.$$

The temperature T is constant by hypothesis, so it comes outside the integral sign. The heat term TS is the part of the system's internal energy that arises from heat flow as the system evolves reversibly from the initial to the final states. If we were to reverse this process, then the amount of mechanical work that we could get out of the system would be W, the remainder of the energy change occurring through the transfer of heat.

The Helmholtz free energy would then be the amount of work that we could derive from this transformation of the system:

HELMHOLTZ FREE ENERGY

$$F = U - TS$$

Because F is the quantity of energy that is *free* to be used to accomplish work (the remainder of the internal energy being committed to heat transfer for the situation in which temperature is maintained as constant), it is called *free energy*.

The free energy is useful for calculating values of other variables associated with the system. For the simple gas, for example,

$$p = -\left(\frac{\partial F}{\partial V}\right)_S = -\left(\frac{\partial U}{\partial V}\right)_S$$

and

$$S = -\left(\frac{\partial F}{\partial T}\right)_V,$$

as can be seen from the definition of F and $p = -(\partial U/\partial V)_V$.

Here is an interesting thing about the free energy. It is the Legendre transformation of the total energy. Recall the use of Legendre transformations in chapter 2 (page 66). The HAMILTONIAN H is the Legendre transformation of the Lagrangian L:

$$H(p, q) = p\dot{q} - L(\dot{q}, q)$$

The effect of this transformation is to define a new function (H) that has the same general properties of the starting function (L) but depends on a different variable $p = \partial L/\partial \dot{q}$ instead of $\dot{q}$. If we write the negative free energy as

$$-F(T, V) = TS - U(S, V),$$

the following correspondence makes the parallel between the two cases evident:

$$H \Leftrightarrow -F$$

$$L \Leftrightarrow U$$

$$q \Leftrightarrow V$$

$$\dot{q} \Leftrightarrow S$$

$$p = \left(\frac{\partial L}{\partial \dot{q}}\right)_q \Leftrightarrow T = \left(\frac{\partial U}{\partial S}\right)_V$$

The Helmholtz free energy, then, results from the Legendre transformation that replaces the extensive variable S with its corresponding intensive variable $T = \partial U/\partial S$, a transformation that is most useful in cases for which the temperature T is constant.

Other Legendre transformations, ones that replace other variables, are equally possible. Two are *enthalpy* and the *Gibbs free energy*:

ENTHALPY

$$H(S, p) = U(S, V) + pV$$

GIBBS FREE ENERGY

$$G(T, p) = U - TS + pV$$

Note that the Gibbs free energy arises from a double Legrendre transformation and that it corresponds to the work that can be derived from a system when it evolves reversibly under conditions of constant temperature and constant pressure. Many chemical experiments are carried out under these conditions (room temperature and atmospheric pressure, for example), so the Gibbs free energy is particularly useful for chemical thermodynamics. The enthalpy is most useful as a surrogate for the total energy under conditions of constant pressure.

During the second half of the nineteenth century, these two laws of thermodynamics proved to be powerful in a variety of applications. The goal of statistical physics has always been to provide a mechanistic explanation for them. In the last half of the nineteenth century, Clausius, Maxwell, and L. E. Boltzmann (1844–1906) developed a statistical interpretation of thermodynamics that forms the basis for modern statistical physics.

Equilibrium Statistical Mechanics

The ideal-gas law had been known since the seventeenth century. Scientists from Newton on tried to explain the springiness of air in terms of constituent atoms, and Daniel Bernoulli devised quite a modern derivation of the ideal-gas law from a molecular view in the eighteenth century. To connect the theory of heat, the thermodynamics of the preceding section, with the main physical theory, mechanics, and thus to complete the unification, a mechanical interpretation of gas behavior, and therefore of the thermal properties of gas, was required. The natural evolution of the subject thus was first to develop the kinetic theory of gases and then to generalize this to other systems. Clausius, Maxwell, and Boltzmann achieved this unification.

The key to a statistical interpretation of the behavior of a gas composed of many particles is the notion that what we actually observe when we measure the properties of the gas is an average and that our physical laws relate not to individual entities but to average quantities. Pressure, for example, is not a property of any *thing* but rather an average effect produced by a large population of entities. This was a really new idea for a physics that had formulated its laws always in terms of the behavior of specific objects and their interactions. Interestingly, an important root of this idea came not from physics but from writings on the social sciences by the Belgian statistician A. Quetelet (1796–1874) in the second third of the nineteenth century. Statistical data indicated that crime and suicide rates were constant over time and between countries, and so one could formulate laws of a "social physics" that related to these average quantities. Maxwell was much taken by these observations when he was a student, and they must certainly have played a role in his thinking about the kinetic theory of gases.

The following treatment of statistical mechanics does not follow its historical evolution but instead provides a slightly more modern development, one that hypothetically could have happened in the nineteenth century.

Suppose that the system we study (think of a gas) is made up of many constituent particles that act independently or with correlations. The energy U of the preceding treatment is an average over time or—equivalently, we assume—over a large collection of identical systems. The system has a variety of microscopic states, or molecular complexions, numbered $1, 2, \ldots, j, \ldots$, and each of these has an associated probability P_j and a specific energy E_j, which is the energy of the state j found by calculating the contributions of all of the constituent particles and their interactions. The thermodynamic internal energy U we used earlier is, we assume, just the average energy:

$$U = \sum_j P_j E_j$$

The *central assumption* of statistical mechanics is that all configurations of the system with the same energy have the same probability of occurring. This assumption will permit us to calculate the probabilities associated with particular states described above.

The Boltzmann relation and the partition function

The fundamental equation that describes the system is the *Boltzmann Relation*:

BOLTZMANN RELATION

$$P_j = \frac{e^{-\beta E_j}}{Z}$$

Here P_j is the probability of observing the jth state with energy E_j, and the quantity Z, which normalizes the probabilities so they sum to 1, is the *partition function*:

PARTITION FUNCTION

$$Z \equiv \sum_j e^{-\beta E_j}$$

And β is a constant whose value we will establish later. The partition function is of particular importance because it provides the key, as will be seen below, for linking statistical mechanics to thermodynamics.

Our development now proceeds in two steps: first is the derivation of the Boltzmann relation, and second is the identification of statistical mechanical quantities with those in thermodynamics.

The goal is to calculate the probability P_j of finding the system in state j—this state has one specific molecular configuration—with an associated energy E_j. To calculate the probability P_j, we consider a small cylinder of gas with a specified volume in thermal contact with a large heat reservoir. Because of the random, complicated movements of the constituent molecules, the state of the gas cylinder and the reservoir will change from instant to instant, but we consider the system only when the gas cylinder happens to be in state j. The energy of the entire system (gas cylinder plus reservoir) is E, and we suppose this to be fixed, because the larger system is completely isolated from its surroundings. The cylinder and reservoir interact only through the interchange of heat, so the reservoir has energy $E_R = E - E_j$ when the cylinder is in state j. Whenever we catch the gas cylinder in one specific microscopic state j with energy E_j, the reservoir can be in any of a large number of possible states subject only to the requirement that they each have energy $E - E_j$. Since every possible mutually exclusive molecular arrangement has the same probability (the central assumption, given above), the probability that the cylinder is in a specified state j is determined by the probability that the reservoir has energy $E - E_j$, and this is proportional to the number of possible molecular arrangements of the reservoir that have energy $E - E_j$ (remember that the probability that one among N mutually exclusive events occurs is Np if each of these events has probability p). The number of distinct states with energy $E - E_j$ is defined to be $\Omega(E - E_j)$, so $P_j \sim \Omega(E - E_j)$.

Now make use of the fact that the reservoir is arbitrarily large, so E_R is arbitrarily close to E. Take the logarithm of $P_j \sim \Omega(E - E_j)$ to make it a more slowly varying function of its argument, and expand around the total energy (remember that energy is determined only up to an additive constant):

$$\ln P_j = \ln \Omega(E - E_j) = \ln \Omega(E) - \beta E_j + \cdots,$$

with $\beta = (\partial \ln \Omega / \partial E)_E$. Since the system has an arbitrary total energy, the remaining terms in the sum can be made as small as required by making the reservoir large enough in comparison with the small cylinder. This means the first two terms in the sum provide an accurate approximation, so

$$P_j = \frac{e^{-\beta E_j}}{Z},$$

where Z is the required proportionality constant necessary to normalize the sum of the probabilities to unity:

$$\sum_j P_j = \sum_j \frac{e^{-\beta E_j}}{Z} = \frac{1}{Z} \sum_j e^{-\beta E_j} = 1,$$

so

$$Z = \sum_j e^{-\beta E_j},$$

the partition function.

The link to thermodynamics

The first step in linking these relations to thermodynamics is to calculate the total energy U. By the assumption above that the thermodynamic U is just the average value over all possible states of the system, we have

$$\begin{aligned}\frac{\partial \ln Z}{\partial \beta} &= \frac{1}{Z}\frac{\partial Z}{\partial \beta} \\ &= \frac{1}{Z}\frac{\partial}{\partial \beta}\sum_j e^{-\beta E_j} \\ &= \frac{-1}{Z}\sum_j E_j e^{-\beta E_j} \\ &= -\sum_j E_j P_j \\ &= -U.\end{aligned}$$

Thus the internal energy is easily calculated from the partition function.

The next step is to calculate the pressure p in the gas cylinder. Here recall that the energy of the jth, state is a function of the volume V of the cylinder. The pressure, then, is

$$\begin{aligned}\frac{\partial \ln Z}{\partial V} &= \frac{1}{Z}\sum_j \frac{\partial}{\partial V} e^{-\beta E_j(V)} \\ &= \frac{-\beta}{Z}\sum_j \frac{\partial E_j(V)}{\partial V} e^{-\beta E_j} \\ &= \beta \sum_j p_j P_j \\ &= \beta p,\end{aligned}$$

where $p_j = -(\partial E_j/\partial V)$ is the pressure associated with the jth state and p is the macroscopic (average) pressure of thermodynamics.

We now have the tools for completing the identification of thermodynamic variables. From the preceding, we can see that $\ln Z(\beta, V)$ is a function of both β and V. Look at the differential of this quantity:

$$
\begin{aligned}
d \ln Z &= \frac{\partial \ln Z}{\partial \beta} d\beta + \frac{\partial \ln Z}{\partial V} dV \\
&= -U\, d\beta + \beta p\, dV \\
&= -d(\beta U) + \beta\, dU + \beta p\, dV \\
&= -d(\beta U) + \beta\underbrace{(dU + p\, dV)}_{T\, dS}
\end{aligned}
$$

Thus,

$$TdS = \frac{1}{\beta} d(\ln Z + \beta U).$$

For this to be true, we must have $1/\beta = kT$ and $dS/k = d(\ln Z + \beta U) = d(\ln Z + U/kT)$ for some constant k, or, up to an additive constant, $S = k \ln Z + U/T$. This means that

$$kT \ln Z = TS - U = -F,$$

where F is the HELMHOLTZ FREE ENERGY, defined in the section on thermodynamics. We thus have the following identification:

$$\beta = \frac{1}{kT}$$

$$-kT \ln Z = F$$

The constant k is known as *Boltzmann's constant.*

A key observation: Once the preceding identification has been made, we can calculate all the thermodynamic properties of a system from the partition function Z because of the central role played by the free energy F in thermodynamics. Thus the partition function is the key for the statistical explanation of thermodynamics.

Finally, we turn to a statistical mechanical representation of entropy. From the relation above,

$$S = k \ln Z + \frac{U}{T},$$

we know that

$$
\begin{aligned}
S &= k \ln Z + k\beta U \\
&= k \ln Z + k \sum_j \beta E_j P_j \\
&= k\left(\ln Z - \sum_j P_j(\ln Z + \ln P_j) \right) \\
&= -k\sum_j P_j \ln P_j
\end{aligned}
$$

(use $\beta = 1/kT$). Note that in the third step we used the relationship $ZP_j = e^{-\beta E_j}$, hence $-\beta E_j = \ln(ZP_j) = \ln Z + \ln P$, and the fact that $\sum_j P_j = 1$. The *entropy relationship* is of central importance as it links the entropy to the probability of the various states:

ENTROPY RELATIONSHIP

$$S = -k \sum_j P_j \ln P_j$$

Entropy can be represented in yet another way. To see this, we must reexamine the partition function,

$$Z = \sum_j e^{-\beta E_j},$$

where the summation is taken over all of the possible microstates j, each with their associated energy E_j. Now divide the possible range of energies for the system into a number of energy bands E_r, each with a width ΔE. The energy interval ΔE is as small as we might require. In general, a large number of the microstates j might fall into the same energy band. Let the number of microstates with energy E_j that fall into the rth band (that is, all states such that $E_r \leq E_j < E_r + \Delta E$) be $\Omega(E_r)$. The partition function can be rearranged to give

$$Z = \sum_j e^{-\beta E_j} = \sum_r \Omega(E_r) e^{-\beta E_r}.$$

This just regroups terms so that ones that fall into the same energy band are together. Now, for a large class of systems, the energy never fluctuates much from its mean value U. For such a system, $\Omega(U)$ will dominate the sum, so, to a good approximation (how good depends on the size of the energy fluctuations), Z is given by just a single term:

$$Z \approx \Omega(U) e^{-\beta U},$$

or

$$\ln Z = \ln \Omega(U) - \beta U.$$

Use $\beta = 1/kT$ and rearrange this last equation to get

$$k \ln Z + \frac{U}{T} = k \ln \Omega.$$

From the definition of $F = U - TS$ and the identification of $F = -kT \ln Z$, we can recognize the left side of the equation above as $S = (-F + U)/T = k \ln Z + U/T$. We thus have the entropy relation that appears on Boltzmann's gravestone (except that Boltzmann used W instead of the Greek equivalent Ω):

ON BOLTZMANN'S
GRAVESTONE

$$S = k \ln \Omega$$

The entropy thus is related to the number of states the system can occupy at its mean energy.

Nonequilibrium Statistical Mechanics*

The preceding discussion was limited to systems at equilibrium, but the approach to equilibrium is, of course, often more interesting. This section presents the fundamental equations used for the study of this problem.

Consider a system that consists of N particles with positions $\mathbf{q}_1, \mathbf{q}_2, \ldots, \mathbf{q}_N$ and momenta $\mathbf{p}_1, \mathbf{p}_2, \ldots, \mathbf{p}_N$. The state of this system can thus be specified as a point $\mathbf{x} = (\mathbf{q}_1, \mathbf{p}_1, \mathbf{q}_2, \mathbf{p}_2, \ldots, \mathbf{q}_N, \mathbf{p}_N)$ in a $6N$-dimensional space called *phase space*. The movement of the point can be calculated for all time from the $6N$ Hamilton's equations, given the initial positions and momenta of all of the particles. Generally, however, the system could have reached a small region in phase space from a large number of different starting points consistent with supplied constraints, like the system's energy. Picture, then, an ensemble of systems that start out from all possible initial conditions (consistent with constraints), and follow this "fluid" of points, one for each system in the ensemble, as they move through phase space. The density of systems at a point $\mathbf{x}$ is denoted by $\omega(\mathbf{x})$.

Now suppose that $A(\mathbf{x})$ is some property of systems at $\mathbf{x}$ (think of the pressure associated with a gas whose state is specified by $\mathbf{x}$). The *fundamental postulate* is that the value $\langle A \rangle$ for the property $A(\mathbf{x})$ that would be observed in an experiment (for example, if one measured the pressure of an actual gas, which can be in many possible states consistent with such constraints as a specified total energy) is the average value given by

$$\langle A \rangle = \int A(\mathbf{x})\rho(\mathbf{x})\,d\mathbf{x},$$

where $\rho(\mathbf{x})$, the probability density, is defined as

$$\rho(\mathbf{x}) = \frac{\omega(\mathbf{x})}{\int \omega(\mathbf{x})\,d\mathbf{x}}.$$

These integrals are taken over all of space, and I have used the shorthand notation

$$\int [\cdot]\,d\mathbf{x} \equiv \int \cdots \int [\cdot]\,d^{6N}x.$$

That is, the density $\omega(\mathbf{x})$ of points at position $\mathbf{x}$ in phase space determines the probability of finding a system in the state that corresponds to that position $\mathbf{x}$. This postulate replaces the one used earlier that all states with the same energy (that is, all positions $\mathbf{x}$ that have the same energy) are equally likely. Note that both ω and ρ are functions of time as well as space because the density of points at any location of phase space will, in general, change as the systems evolve.

The goal now is to develop an equation that governs the evolution of $\rho(\mathbf{x}, t)$ over time. The continuity equation for movement of the "fluid" in the $6N$-dimensional space will provide the required description.

Define the $6N$-dimensional divergence operator:

$$\nabla \cdot \equiv \left(\frac{\partial}{\partial q_1}, \frac{\partial}{\partial p_1}, \ldots, \frac{\partial}{\partial q_N}, \frac{\partial}{\partial p_N} \right)$$

Because systems must move continuously in phase space with their total number conserved, for any closed region R the rate at which the number of points inside R changes must equal the rate of flow of points across the region's surface:

$$\underbrace{\frac{\partial}{\partial t}\int_V \omega(\mathbf{x},t)\,d\mathbf{x}}_{\text{rate of change of contents}} = \underbrace{-\int_S \omega(\mathbf{x},t)\dot{\mathbf{x}}\cdot d\mathbf{S}}_{\text{inflow through surface}}$$

Here the left integral is over the volume of R, and the right integral is over its surface, with a surface element $d\mathbf{S}$ (the normal of this vector points outward from the surface); $\dot{\mathbf{x}}$ is the velocity of points at $\mathbf{x}$. Use the DIVERGENCE THEOREM (generalized to $6N$ dimensions) to get

$$\int_V \frac{\partial \omega}{\partial t}\,d\mathbf{x} = -\int_S \omega(\mathbf{x},t)\dot{\mathbf{x}}\cdot d\mathbf{S} = -\int_V \nabla\cdot(\omega\dot{\mathbf{x}})\,d\mathbf{x}.$$

Because the region R is arbitrary, it must be that

$$\frac{\partial \omega}{\partial t} = -\nabla\cdot(\omega\dot{\mathbf{x}})$$

$$= \nabla\omega\cdot\dot{\mathbf{x}} - \omega\nabla\cdot\dot{\mathbf{x}}.$$

The right side of this equation, in terms of the components of $\mathbf{x}$, is

$$-\nabla\omega\cdot\dot{\mathbf{x}} - \omega\nabla\cdot\dot{\mathbf{x}} = -\sum_k\left(\frac{\partial\omega}{\partial q_k}\dot{q}_k + \frac{\partial\omega}{\partial p_k}\dot{p}_k\right) - \omega\sum_k\left(\frac{\partial\dot{p}_k}{\partial p_k} + \frac{\partial\dot{q}_k}{\partial q_k}\right).$$

Now use HAMILTON'S EQUATIONS $\dot{q}_k = \partial H/\partial p_k$ and $\dot{p}_k = -\partial H/\partial q_k$ to eliminate $\dot{q}_k$ and $\dot{p}_k$ from the preceding hence to get

$$\frac{\partial\omega}{\partial t} = -\sum_k\left(\frac{\partial\omega}{\partial q_k}\frac{\partial H}{\partial p_k} - \frac{\partial\omega}{\partial p_k}\frac{\partial H}{\partial q_k}\right) - \omega\sum_k\overbrace{\left(\frac{\partial^2 H}{\partial p_k\partial q_k} - \frac{\partial^2 H}{\partial p_k\partial q_k}\right)}^{\text{vanishes}}$$

$$= \{H,\omega\},$$

where $\{H,\omega\}$ is, by definition, the POISSON BRACKET. Since $\rho \sim \omega$, this same equation holds for the probability density ρ. Thus we have the *Liouville equations*:

LIOUVILLE EQUATIONS

$$\frac{\partial\omega}{\partial t} = \{H,\omega\}$$

$$\frac{\partial\rho}{\partial t} = \{H,\rho\}$$

The Liouville equation can be solved formally to give the *Liouville evolution equations*:

LIOUVILLE EVOLUTION EQUATIONS

$$\rho(t) = e^{t\{H,\ \}}\rho(0),$$

which makes use of the *Liouville operator*:

$$\{H,\ \} \equiv \sum_k \left(\frac{\partial H}{\partial q_k}\frac{\partial}{\partial p_k} - \frac{\partial H}{\partial p_k}\frac{\partial}{\partial q_k} \right)$$

(See page 14 for the meaning of an operator that appears in an exponential.) These equations govern the movement of points in phase space and thus specify the approach of a system to its equilibrium configuration.

At equilibrium, $\partial\rho/\partial t = 0$, so $\{H, \rho\} = 0$. In the preceding section, the probability of an equilibrium state was identified as a function of the energy, so the relation $\{H, \rho(E)\} = 0$ needs to be verified. It turns out that this relation holds for any function ρ that depends only on the energy, E. To see this, write

$$\begin{aligned}\{H, \rho(E)\} &= \sum_k \left(\frac{\partial H}{\partial q_k}\frac{\partial \rho(E)}{\partial p_k} - \frac{\partial H}{\partial p_k}\frac{\partial \rho(E)}{\partial q_k} \right)\\ &= \sum_k \left(\frac{\partial H}{\partial q_k}\frac{d\rho}{dE}\frac{\partial E}{\partial p_k} - \frac{\partial H}{\partial p_k}\frac{d\rho}{dE}\frac{\partial E}{\partial q_k} \right)\\ &= \frac{d\rho}{dE}\sum_k \left(\frac{\partial H}{\partial q_k}\frac{\partial E}{\partial p_k} - \frac{\partial H}{\partial p_k}\frac{\partial E}{\partial q_k} \right)\\ &= \frac{d\rho}{dE}\underbrace{\{H, E\}}_{\text{vanishes}}.\end{aligned}$$

Because E is a constant of the motion, $\{H, E\} = 0$, so $\{H, \rho(E)\} = 0$ whenever ρ is an equilibrium probability density (see page 68).

The equilibrium density $\rho(E)$ can be determined by the perturbation argument used in the preceding section (see page 154) to give the BOLTZMANN RELATION,

$$\omega(\mathbf{x}) = e^{-\beta E(\mathbf{x})} \quad \text{and} \quad \rho(\mathbf{x}) = \frac{e^{-\beta E(\mathbf{x})}}{Z},$$

with the PARTITION FUNCTION

$$Z = \int \omega(\mathbf{x})\, d\mathbf{x}.$$

Of course, all of the thermodynamic quantities are derived from the partition function, as was described earlier.

The treatment given here is an alternative to the equilibrium statistical mechanics developed in the preceding section, but this argument is easily generalized to calculate the approach to equilibrium (in principle, if not in practice). Given an initial distribution, one simply uses the Liouville evolution equation. The problem, of course, is that the required calculations usually cannot be performed. Most frequently, then, one resorts to calculations of how systems relax from one equilibrium state to another neighboring equilibrium state when the system is perturbed slightly by applying an external driving force.

Suppose that we have a system with an equilibrium probability density ρ_0 that corresponds to a particular Hamiltonian H_0, and then at time $t = 0$ we switch on a perturbation to this Hamiltonian λH_{p}, so the new Hamiltonian is $H = H_0 + \lambda H_{\mathrm{p}}$. Call Liouville operator

$$M \equiv \{H, \quad\}.$$

So

$$M = \{H_0 + \lambda H_{\mathrm{p}}, \quad\} \equiv M_0 + \lambda M_{\mathrm{p}}.$$

The constant λ sets the strength of the perturbation. The Liouville equation for this situation is thus

$$\frac{d\rho}{\partial t} = (M_0 + \lambda M_{\mathrm{p}})\rho,$$

with the initial condition $\rho(0) = \rho_0$ (which we know). If a system starts at its final equilibrium distribution, it does not evolve further. That is, according to the Liouville evolution equation,

$$\rho_0 = e^{M_0 t}\rho_0.$$

The solution to the Liouville equation is thus

$$\rho(t) = e^{(M_0 + \lambda M_{\mathrm{p}})t}\rho_0 = e^{\lambda M_{\mathrm{p}} t} e^{M_0 t}\rho_0 = e^{\lambda M_{\mathrm{p}} t}\rho_0.$$

If λ is sufficiently small, this equation is approximately

$$\rho(t) = \rho_0 + \lambda M_{\mathrm{p}} t \rho_0$$

(use $e^{\lambda M_{\mathrm{p}} t} \approx 1 + \lambda M_{\mathrm{p}} t$). Because M_{p} is a linear operator, this approach can be valid when the response of the system is within a linear range. Variations of this method are useful for examining the behavior of systems near equilibrium.

This approach depends on perturbations of a system around equilibrium. An alternative method connects a quantity that can be calculated for equilibrium systems, the correlation function of some variable, with relaxations towards equilibrium, a nonequilibrium process. The idea is as follows. According to the Liouville evolution equation, the development of the probability $\rho(t)$ depends only on its initial state and not on any path through which that state developed. That is, statistical mechanical systems evolve according to Markov processes insofar as they can be adequately represented by a random process. If $A(\mathbf{x})$ is some measurable property associated with state $\mathbf{x}$ (and we assume a mean of 0 and a stationary system), the covariance $C(t)$ for fluctuations in A is, by definition, given by the equation

$$C(t) = \int\int d\mathbf{x} d\mathbf{x}' \, A(\mathbf{x}) A(\mathbf{x}') \rho(\mathbf{x}, t; \mathbf{x}', 0),$$

where $\rho(\mathbf{x}, t; \mathbf{x}', 0)$ is the joint probability of finding state $\mathbf{x}$ at time t and state $\mathbf{x}'$ at time 0. The joint probability can be written as

$$\rho(\mathbf{x}, t; \mathbf{x}', 0) = \rho(\mathbf{x}') \rho(\mathbf{x}, t | \mathbf{x}'),$$

where $\rho(\mathbf{x}')$ is the initial probability density and $\rho(\mathbf{x}, t|\mathbf{x}')$ is the conditional probability of finding the system in state $\mathbf{x}$ at time t, given that it started in state $\mathbf{x}'$. The covariance thus can be written as

$$C(t) = \int d\mathbf{x}' \, A(\mathbf{x}') \rho(\mathbf{x}') M(\mathbf{x}', t)$$

for the conditional mean $M(\mathbf{x}', t)$, which is defined as follows:

$$M(\mathbf{x}', t) \equiv \int d\mathbf{x} \, A(\mathbf{x}) \rho(\mathbf{x}, t | \mathbf{x}')$$

According the mean-value theorem, for some $\mathbf{x}' = \bar{\mathbf{x}}$ we can write the covariance function as

$$C(t) = A(\bar{\mathbf{x}})M(\bar{\mathbf{x}}, t) \int d\mathbf{x}' \rho(\mathbf{x}') = A(\bar{\mathbf{x}})M(\bar{\mathbf{x}}, t).$$

If the system is Markovian, however, the path followed to a particular state has no influence on the way the system relaxes from that state. Suppose that we instantly drive the system (with a delta function) away from equilibrium into an initial state characterized by probability density $\rho_a(\mathbf{x}')$. Then the average behavior $G(t)$ that describes the relaxation from that state is

$$\begin{aligned} G(t) &= \int\int d\mathbf{x}d\mathbf{x}'\rho_a(\mathbf{x}')\rho(\mathbf{x}, t|\mathbf{x}')A(\mathbf{x}) \\ &= \int d\mathbf{x}' \rho(\mathbf{x}')M(\mathbf{x}', t) \\ &= M(\bar{\mathbf{x}}, t). \end{aligned}$$

The last step used the mean-value theorem, and $M(\bar{\mathbf{x}}, 0)$ is the starting point of the perturbed system. Because $G(t)$ is the response to delta-function driving, $G(t)$ is the Green's function if the system is within a linear response range. Hence, for a Markovian linear system, the Green's function and the covariance function are proportional, so we have the *fluctuation-dissipation theorem*:

FLUCTUATION-DISSIPATION THEOREM

$$C(t) \sim G(t)$$

This theorem is important because it connects nonequilibrium behavior, a perturbed system's relaxing back to equilibrium, to a function that can be calculated for the equilibrium state, the correlation function.

Quantum Statistical Mechanics*

The earlier development of equilibrium statistical mechanics (the section starting on page 151) did not make explicit use of classical mechanics but rather just assumed the existence of various microscopic states with associated energy levels. Thus the results of that section should hold either for classical or quantum systems. In the preceding section on nonequilibrium statistical mechanics, however, the formalism did explicitly use classical mechanics, embodied in the Liouville equation, and thus the results there are restricted to classical systems. The goal here is to generalize the considerations of the preceding section to quantum-mechanical systems.

The phase space of classical statistical mechanics for a system of N particles is $6N$-dimensional, with a state of the system being specified by the positions and momenta of all of the particles. Because of the Heisenberg uncertainty principle, a quantum system cannot simultaneously specify both positions and momenta, so the phase space is replaced by a *configuration space* that has as components of its position vector $\mathbf{x} = (\mathbf{x}_1, \mathbf{x}_2, \ldots, \mathbf{x}_N)$ the arguments of the system's wave function $\psi(\mathbf{x}_1, \mathbf{x}_2, \ldots, \mathbf{x}_N)$. For N particles, this is usually the $3N$ positions.

Our knowledge of a macroscopic state of a system—its pressure, for example—is consistent with many possible individual microscopic realizations of that system. In the classical case, each possible realization of the macroscopic state is represented by a single system whose complete state is known, so the macroscopic state is associated with an ensemble of individual, completely specified systems. At any specific time, each of these individual systems occupies a single point in phase space, and the probability of finding a macroscopic state is given by the number of points in phase space that are consistent with the macroscopic state.

For the quantum-mechanical case, however, each of these individual systems is represented by a wave function, and the probability of finding a particular configuration, that is, a particular point $\mathbf{x}$ in configuration space, now depends on two things: the first is the number of elementary systems (that is, basis vectors) present in the ensemble, and the second is the probability that each of these elementary systems is found at position $\mathbf{x}$ in configuration space. This second factor is peculiar to the quantum-mechanical case. The density of systems in the ensemble that are in the vicinity of $\mathbf{x}$ was called $\omega(\mathbf{x})$ in the classical case; the fact that any system, represented by its wave function, has some probability to be at $\mathbf{x}$ is new for quantum mechanics. The first step in the development of quantum statistical mechanics, then, will be to generalize the density function ω and the probability density ρ from the classical to the quantum-mechanical situations. I will then develop quantum statistical mechanics as I did classical mechanics: by defining the average value of a quantity A associated with a position $\mathbf{x}$ in configuration space, finding the quantum-mechanical analog of the Liouville equation, and finally considering the equilibrium state for quantum systems.

The most convenient approach is to start by identifying the energy eigenfunctions for the configuration space. The basis vectors for the wave functions that describe configuration space are ϕ_n, where $\mathbf{H}\phi_n = E_n\phi_n$ ($\mathbf{H}$ is the Hamiltonian operator, and E_n is the energy level, that is, the eigenvalue, that corresponds to ϕ_n). Note that the $\phi_n(\mathbf{x})$ are constructed to be consistent with the macroscopic constraints on the system we are describing and are functions of the variables, represented by the vector $\mathbf{x}$, that are required for specifying a position in configuration space. Each of the eigenfunctions ϕ_n can be thought of as the analog of an individual phase-space point (i.e., system) in the classical case.

A specific representative system k of the sort we consider (for example, the system whose pressure we are trying to calculate) might have a wave function $\psi^{(k)}$. Generally, a large number of different wave functions (systems) might be consistent with the macroscopic properties we wish to characterize. Now define a wave function for the entire ensemble as $\psi(\mathbf{x}) \equiv \sum_k \psi^{(k)}$, where ψ has been normalized and the sum is taken over all wave functions consistent with the provided constraints (like a specified volume and energy range). The ensemble wave function ψ can be represented as a superposition of the elementary wave functions:

$$\psi(\mathbf{x}) = \sum_n \overline{\langle \psi^{(k)} | \phi_n \rangle} \phi_n,$$

where $\overline{\langle \psi^{(k)} | \phi_n \rangle}$ is the average taken over all of the k possible wave functions of representative systems. Now define the number of elementary systems in the ensemble with wave function ϕ_n to be

$$\omega_n = |\overline{\langle \psi^{(k)} | \phi_n \rangle}|^2.$$

With this definition, the density of elementary systems at location $\mathbf{x}$ in configuration space is specified by the density function $\omega(\mathbf{x})$, defined as

$$\omega(\mathbf{x}) = \sum_n \phi_n^*(\mathbf{x}) \omega_n \phi_n(\mathbf{x}) = \sum_n \omega_n |\phi_n(\mathbf{x})|^2 = \sum_n \omega_n P_n(\mathbf{x}).$$

Here $P_n(\mathbf{x}) \equiv |\phi_n(\mathbf{x})|^2$ is the probability of finding an elementary system with wave function $\phi_n(\mathbf{x})$ at $\mathbf{x}$ in configuration space, and ω_n is the number of systems with wave function $\phi_n(\mathbf{x})$ in the ensemble. The density function $\omega(\mathbf{x})$ is the quantum-mechanical analog of the classical density function in the preceding section.

For some problems in quantum statistical mechanics, a slight generalization of the preceding definition of the density function is useful. Consider two points $\mathbf{x}$ and $\mathbf{x}'$ in configuration space, and define the *density function* as follows:

DENSITY FUNCTION

$$\omega(\mathbf{x}, \mathbf{x}') = \sum_n \phi_n^*(\mathbf{x})\omega_n\phi_n(\mathbf{x}')$$

The density function $\omega(\mathbf{x})$ above depends on two variables now $\omega(\mathbf{x}, \mathbf{x}')$. As for the classical case, we define the normalized density function to be the *probability density function*:

PROBABILITY DENSITY FUNCTION

$$\rho(\mathbf{x}, \mathbf{x}') = \frac{\omega(\mathbf{x}, \mathbf{x}')}{Z}$$

This has the following *quantum partition function*:

QUANTUM PARTITION FUNCTION

$$Z \equiv \int d\mathbf{x}\, \omega(\mathbf{x}, \mathbf{x})$$

Note that the argument of the density function in the integral is $(\mathbf{x}, \mathbf{x})$, and that I have used the convention $\int [\cdot]\, d\mathbf{x}$ established in the preceding section for a multidimensional integral over the permissible regions of configuration space.

Because operators in quantum mechanics play the role of observable quantities in classical physics, we need to associate an operator with the density function defined above. The natural definition is the *density matrix operator*:

DENSITY MATRIX (OPERATOR)

$$\omega \equiv \sum_n |\phi_n\rangle \omega_n \langle\phi_n|$$

This operator will act on a system in a pure state ϕ_k to give the density ω_k of states in the ensemble with the energy E_k:

$$\langle\phi_j|\omega|\phi_k\rangle = \sum_n \langle\phi_j|\phi_n\rangle\omega_n\langle\phi_n|\phi_k\rangle = \delta_{jk}\omega_k,$$

where δ_{jk} is the Kronecker delta function. Thus the density operator can be used to specify the number of representatives of an elementary system that the ensemble contains; this will be useful later for following the evolution of the ensemble over time and for the equilibrium case, to which Boltzmann's relationship applies. If the density operator acts on wave functions $|\mathbf{x}\rangle$ that specify a particular position in configuration space (that is, delta-function wave functions), then the density operator can be used to represent the density function described earlier:

$$\omega(\mathbf{x}, \mathbf{x}') = \langle \mathbf{x} | \omega | \mathbf{x}' \rangle$$

Of course, the density function and the density operator were designed specifically to have this property. The probability density matrix (operator) ρ is just the normalized version of the density matrix just discussed.

For the classical system, the observed value $\langle A \rangle$ of any property of the system $A(\mathbf{x})$ is the average

$$\langle A \rangle = \int d\mathbf{x}\, A(\mathbf{x}) \rho(\mathbf{x}).$$

For a quantum system, the corresponding equation, for an operator $\mathbf{A}$ that represents an observable quantity a, is

$$\langle A \rangle = \sum_n \langle \phi_n | \mathbf{A} \rho | \phi_n \rangle = \sum_n \rho_n \langle \phi_n | \mathbf{A} | \phi_n \rangle = \sum_n \rho_n a_n,$$

where $a_n = \langle \phi_n | \mathbf{A} | \phi_n \rangle$ is the value of the observable a for a pure system with energy E_n, and ρ_n is the probability that the pure system represented by the wave function ϕ_n is in the ensemble. By definition of the *trace* of a matrix (operator),

$$\langle A \rangle = \sum_n \langle \phi_n | \mathbf{A} \rho | \phi_n \rangle \equiv \mathrm{Tr}[\mathbf{A}\rho].$$

Another way to write this is

$$\langle A \rangle = \int d\mathbf{x} \sum_n \phi_n^*(\mathbf{x}) \rho_n \mathbf{A} \phi_n(\mathbf{x}) \equiv \mathrm{Tr}[\mathbf{A}\rho].$$

We turn now to the time evolution of quantum mechanical ensembles. An ensemble starts in some specified state $\psi(0)$ with an associated density function $\omega(x, x')$, but of course it evolves according to the evolution equation

$$\psi(t) = \exp\left(\frac{-i}{\hbar} \mathbf{H} t\right) \psi(0).$$

As this evolution progresses, the number of elementary systems ϕ_n that must be present in the ensemble to give $\psi(t)$ also will change. The time course of this change, for a Hamiltonian $\mathbf{H}$ that is independent of time, is given by the *density-operator equation of motion*:

DENSITY-OPERATOR EQUATION OF MOTION

$$\frac{d\omega(t)}{dt} = \frac{i}{\hbar}[\mathbf{H}, \omega]$$

(See the OPERATOR EQUATION OF MOTION on page 131.) This is the quantum-mechanical analog of the classical LIOUVILLE EQUATION. Like that equation, it can be used to solve problems involving the relaxation to equilibrium.

For a system at equilibrium, $d\omega/dt = 0$, so $[\mathbf{H}, \omega] = 0$, just as $\{H, \omega\} = 0$ for the classical case. The argument used earlier to arrive at the Boltzmann distribution (see page 153) made no specific use of classical physics, so the argument is equally valid for the quantum-mechanical ensemble. Thus, as before, we have a Boltzmann relation:

QUANTUM BOLTZMANN RELATION

$$\omega_n = e^{-\beta E_n}$$

Here $\beta = 1/kT$. The energy levels for the quantum Boltzmann relation are provided by the quantum-mechanical equations, whereas they were arbitrary for the classical case. For an equilibrium ensemble, the probability-density function $\rho_0(\mathbf{x}, \mathbf{x}')$ is just the normalized density function:

$$\rho_0(\mathbf{x}, \mathbf{x}') = \frac{1}{Z}\sum_n \phi_n^*(\mathbf{x}) e^{-\beta E_n} \phi_n(\mathbf{x}')$$

Here Z, the partition function, is given by

$$Z = \sum_n e^{-\beta E_n}.$$

Alternatively,

$$Z = \int d\mathbf{x}\, \omega(\mathbf{x}, \mathbf{x}),$$

with

$$\omega(\mathbf{x}, \mathbf{x}') = \sum_n \phi_n^*(\mathbf{x}) e^{-\beta E_n} \phi_n(\mathbf{x}').$$

The partition function is most usually represented in terms of the density operator:

$$\omega = \sum_n |\phi_n\rangle e^{-\beta E_n}\langle\phi_n| = e^{-\beta \mathbf{H}} \sum_n |\phi_n\rangle\langle\phi_n| = e^{-\beta \mathbf{H}}$$

The first step made use of the fact that $\mathbf{H}\phi_n = E_n\phi_n$ and the property of operators and their eigenvalues that

$$\sum_n f(\mathbf{H})\phi_n = \sum_n f(E_n)\phi_n$$

(see the SPECTRAL THEOREM). So

$$e^{-\beta E_n}\phi_n = e^{-\beta \mathbf{H}}\phi_n.$$

The second step used the identification of $\sum_n |\phi_n\rangle\langle\phi_n|$ with the identity operator. This representation of the density operator provides the usual version of the *quantum-mechanical partition function*:

QUANTUM-MECHANICAL
PARTITION FUNCTION

$$Z = \mathrm{Tr}[e^{-\beta \mathbf{H}}]$$

This relation depends on the definition of the trace as

$$\mathrm{Tr}[e^{-\beta \mathbf{H}}] \equiv \sum_n \langle\phi_n|e^{-\beta \mathbf{H}}|\phi_n\rangle = \sum_n e^{-\beta E_n}\langle\phi_n|\phi_n\rangle = \sum_n e^{-\beta E_n}.$$

Remember that the $\{\phi_n\}$ are orthonormal.

The density function and the density matrix are constructed to contain the same information, and either can be used to represent the partition function for the equilibrium situation, as described earlier:

$$Z = \int dx\, \omega(\mathbf{x}, \mathbf{x}) = \mathrm{Tr}[\omega] = \mathrm{Tr}[e^{-\beta \mathbf{H}}]$$

$$\omega(\mathbf{x}, \mathbf{x}) = \langle\mathbf{x}|e^{-\beta \mathbf{H}}|\mathbf{x}\rangle$$

Here $|\mathbf{x}\rangle$ is a wave function that definitely places the system at position $\mathbf{x}$ in configuration space. This last expression leads to an interesting formal similarity between statistical mechanics and quantum mechanics that permits the density function to be written as a functional integral, a representation that can often facilitate the formulation and solution of problems. The key to this connection is to notice that the transition amplitude

$K(\mathbf{x}, t; \xi, 0)$ can be written

$$K(\mathbf{x}, t; \xi, 0) = \langle \mathbf{x} | \exp(-it\mathbf{H}/\hbar) | \xi \rangle = \int \mathscr{D}\mathbf{x} \exp\left(\frac{i}{\hbar}\int_0^t dt' \, L(\mathbf{x}, \dot{\mathbf{x}})\right),$$

where $\langle \mathbf{x}|$ and $|\xi\rangle$ are the wave functions for the indicated locations in configuration space and the functional integral is taken over all paths that start at ξ and end at $\mathbf{x}(t)$. The trick is that the expressions for $K(\mathbf{x}, t; \xi, 0) = \langle \mathbf{x} | \exp(-(i/\hbar)t\mathbf{H}) | \xi \rangle$ and $\omega(\mathbf{x},\mathbf{x}) = \langle \mathbf{x} | \exp(-\beta\mathbf{H}) | \mathbf{x} \rangle$ are identical except for two things: first, ξ appears in the transition amplitude instead of $\mathbf{x}$, and second, $it/\hbar$ appears in the transition amplitude instead of β. The idea is to consider the transition amplitudes that start and end in the same states and to let the transition occur in imaginary time it so that the substitution $\beta\hbar = it$ can be made.

First, look at the transition amplitude that starts and ends at $\mathbf{x}$ and occurs in imaginary time:

$$K(\mathbf{x}, it; \mathbf{x}, 0) = \int \mathscr{D}\mathbf{x} \exp\left(\frac{i}{\hbar}\int_0^{it} dt' \, L(\mathbf{x}, \dot{\mathbf{x}})\right)$$

Now let imaginary time $\tau = it'$ run from 0 to it in the integral that appears as argument to the exponential function:

$$\frac{1}{\hbar}\int_0^{it} i \, dt' \, L(\mathbf{x}, d\mathbf{x}/dt) = \frac{1}{\hbar}\int_0^{it} d\tau \, L(\mathbf{x}, i \, d\mathbf{x}/d\tau)$$

Note that on the left $d\mathbf{x}/dt$ appears, whereas on the right $d\mathbf{x}/d\tau = d\mathbf{x}/i\,dt$ is present; therefore, $i\,d\mathbf{x}/d\tau$ must appear as an argument of the Lagrangian L. If $\beta\hbar = it$, then

$$\omega(\mathbf{x},\mathbf{x}) = \langle \mathbf{x} | e^{-\beta\mathbf{H}} | \mathbf{x} \rangle = K(\mathbf{x}, \beta\hbar; \mathbf{x}, 0) = \int \mathscr{D}\mathbf{x} \exp\left(\frac{1}{\hbar}\int_0^{\beta\hbar} d\tau \, L(\mathbf{x}, i \, d\mathbf{x}/d\tau)\right),$$

where $\omega(\mathbf{x},\mathbf{x})$ has been represented by a functional integral. Note that τ plays the role of time but that its physical significance in this context is reciprocal temperature. For the case of a gas consisting of N particles with mass m and a potential energy function $V(\mathbf{x})$, $L(\mathbf{x}, \dot{\mathbf{x}}) = \sum -m\dot{x}_k^2/2 - V(\mathbf{x})$, and the integral in the argument to the exponential function is

$$\frac{1}{\hbar}\int_0^{\beta\hbar} d\tau \, L(\mathbf{x}, i \, d\mathbf{x}/d\tau) = \frac{1}{\hbar}\int_0^{\beta\hbar} d\tau \left(\sum_{k=1}^{N} \frac{-m\dot{x}_k^2}{2} - V(\mathbf{x})\right).$$

The minus sign in the kinetic-energy term arises because $d\mathbf{x}/dt$ is replaced by $i\,d\mathbf{x}/d\tau$ when time is converted into imaginary time. Thus the density function becomes

$$\omega(\mathbf{x},\mathbf{x}) = \int \mathscr{D}\mathbf{x} \exp\left(-\frac{1}{\hbar}\int_0^{\beta\hbar} d\tau \left(\sum_{k=1}^{N} \frac{m\dot{x}_k^2}{2} + V(\mathbf{x})\right)\right),$$

with the functional integral taken over all paths that start and end at $\mathbf{x}$. For most such paths, $V(\mathbf{x})$ will not vary much (in the high temperature limit), so the integral in the argument to the exponential function is

$$-\frac{1}{\hbar}\int_0^{\beta\hbar} d\tau\, V(\mathbf{x}(\tau)) \approx -\beta V(\mathbf{x})$$

($\mathbf{x}$ is the starting and ending location in configuration space), and the density function is approximately

$$\omega(\mathbf{x},\mathbf{x}) \approx \exp(-\beta V(\mathbf{x})) \int \mathscr{D}\mathbf{x} \exp\left(-\frac{1}{\hbar}\int_0^{\beta\hbar} d\tau \sum_{k=1}^{N} \frac{m\dot{x}_k^2}{2}\right).$$

The remaining functional integral can be recognized to be like that for a free particle (moving through imaginary time $1/kT = \beta = it/\hbar$ from $\mathbf{x}$ to $\mathbf{x}$) (see page 108). So we have

$$\int \mathscr{D}\mathbf{x} \exp\left(-\frac{1}{\hbar}\int_0^{\beta\hbar} d\tau \sum_{k=1}^{N} \frac{m\dot{x}_k^2}{2}\right) = \sqrt{\frac{NmkT}{2\pi\hbar^2}}.$$

Thus the density function for this simple case is

$$\omega(\mathbf{x},\mathbf{x}) = \sqrt{\frac{NmkT}{2\pi\hbar^2}} \exp(-V(\mathbf{x})/kT),$$

and the partition function is

$$Z = \sqrt{\frac{NmkT}{2\pi\hbar^2}} \int d\mathbf{x} \exp(-V(\mathbf{x})/kT),$$

which is a standard result for a gas in which the particles interact according to a potential-energy function $V(\mathbf{x})$.

6 Special Relativity

Special relativity is not hard to understand. It is hard to believe. As a consequence of this disbelief, the intuitions one uses for solving problems cannot be trusted, and special relativity gets very hard to apply.

The development of special relativity presented here takes place in two main steps. The first is a presentation of Einstein's postulates and some of the consequences of the idea that light has the same speed in different frames of reference. The second step explores implications of the postulate that physical laws have the same form in any *inertial frame* (defined below).

Einstein's Postulates and First Consequences

In his theory of special relativity, Einstein views space as Euclidean and limits his considerations to inertial frames of reference. An *inertial frame* is a coordinate system that is either fixed or moving linearly at a constant speed relative to the fixed stars. Strictly speaking, an inertial frame should be in a region of space that is free of gravity, but in practice, other frames moving at a constant velocity are used. The idea of an inertial frame is that Newton's law of inertia should hold: bodies in motion continue indefinitely in a straight line unless some force acts.

FIRST POSTULATE The speed of light is the same in all inertial frames.

This is the hard thing to accept. It means that if two observers, one in a fixed and one in a rapidly moving frame of reference, watch the spreading of light from a brief flash, they will see identical behavior of the light in both frames. Suppose, for example, that the light flash occurred at the origin of both frames of reference at the instant that the origin of the moving frame coincided with that of the fixed frame. Suppose further that the moving frame is traveling at, say, half the speed of light. Then an observer in the fixed frame will see the light spread out spherically from the origin with a velocity c. Here is the amazing part: the observer riding in the moving frame of reference (a very fast rocket, for example) will also see the light spread out spherically from the origin of *his* coordinate system with velocity c in all directions. Thus, in the direction the rocket is moving, the light-wave front still moves, relative to the rocket, with speed c instead of the naively expected $c/2$. This postulate leads to strange results, like time going at a different rate in a moving coordinate system. The nonconstancy of time has been verified by measuring the predicted differences in the rate at which subatomic particles decay when they are slowly or rapidly moving.

SECOND POSTULATE The laws of physics should have the same form when they are determined in any inertial frame of reference.

Laws that follow this principle are said to be *covariant*, which means that all of their variables change together (covary) in just the right way to keep their relation unchanged when the coordinates are transformed from a fixed frame of reference to a moving one. For example, the covariance of Newton's law would mean that if $f = ma$ in a fixed frame, then $f' = m'a'$ would hold in a moving (primed) frame. Although perhaps not obvious, this is easier to accept.

The next job is to examine some of the implications of these two postulates. We begin with the first postulate. The goal here is to find a relation between the two reference frames. More specifically, imagine that an observer in a fixed frame F and another observer in a moving frame F' make measurements on the same event. Picture, for example, someone on the ground and someone in an airplane tracking an object (like a bomb) falling from the airplane. To the ground observer, the object will fall to the ground in a long arc, while the observer in the plane (neglecting air-resistance effects) will see the same object falling straight down. We want to be able to translate back and forth between the two reference frames, so that knowing the behavior of the object in one frame will permit predicting precisely what an observer would see in the second frame. It will turn out that the rule relating these two frames will be the *Lorentz transformation*, but before I describe this transformation, we need a definition of the alternative *Galilean* transformation that it replaced.

The Galilean transformation relates events in fixed and moving reference frames

Before Einstein, the relation between a fixed coordinate system (F) and a moving one (F', distinguished by the primes) was the *Galilean transformation*:

GALILEAN TRANSFORMATION

$$x' = x - ut$$

$$t' = t$$

Here x is (say) a particle's position in the fixed coordinate system at time t, and x' is the position of the same particle in the moving coordinate system at time $t' = t$; the second frame moves with a constant velocity u relative to the first. According to this transformation, an object resting at x in F (that is, in F, x is not changing with time) would appear to be moving with a velocity $dx'/dt' = -u$ as viewed by an observer in F'. Picture a train station that is fixed for those waiting for a train (F) but traveling backwards with velocity u (the train's speed) for an observer riding on a train (F') that passes though the station. Here we assume that time is universal, so F and F' can share a clock for their time measurements. Also the coordinate axes are aligned so that $y = y'$ and $z = z'$. Newton's laws were known to be invariant under the Galilean transformation, but this is all so obvious that it is hard to understand why anyone would bother about it.

Here is an illustration of a use for this transformation to show that it leaves Newton's equation unchanged: Let a particle, moving under a force f derivable from potential V, be at position x at time t in F and at x' at $t' = t$ in F'. The same particle motions are being measured in F and F', but the observer doing the measuring is fixed F, and the observer is moving along with the coordinate system (whose velocity is u) in F'. Newton's law in F is

$$f = \partial_x V(x, t) = ma.$$

The Galilean transformation applied to this law gives, for F',

$$f' = \frac{\partial V(x' + ut', t')}{\partial x'} \equiv \frac{\partial V'(x', t')}{\partial x'} = m\frac{d^2(x' + ut')}{dt^2} = m\frac{d^2x'}{dt'^2} = ma'$$

(use the Galilean transformation equations to eliminate x and t in favor of x' and t', note that $\partial x' = \partial x$, and remember that $t' = t$). The form is just the same in F' as in F. So Newton's law is form-invariant (covariant) under the Galilean transformation.

The wave equation for the electromagnetic field (derived from Maxwell's equations; see page 96) is not, however, form-invariant under this transformation. To see this, carry out the Galilean transformation as before on the wave equation

$$\frac{1}{c^2}\frac{\partial^2 U(x, t)}{\partial t^2} = \frac{\partial^2 U(x, t)}{\partial x^2}$$

(for constant c and a waving function $U(x,t)$) to get

$$\frac{1}{c^2}\frac{\partial^2 U(x'+ut',t')}{\partial t'^2}=\frac{\partial^2 U(x'+ut',t')}{\partial x'^2}.$$

We can identify $U'(x',t') \equiv U(x'+ut',t')$, but because x and t are treated symmetrically in the wave equation, but not by the Galilean transformation, the transformed wave equation will no longer generally hold. This is true because the spatial variable on the right appears only in one of the arguments of U, but the temporal variable appears in both arguments on the left. So only for special cases will the spatial derivative have the same result as the temporal one.

Something is wrong, and Einstein, because he already was convinced that light comes in quantal packets, finally decided, despite some suspicions about Maxwell's equations, that the fault lies with the Galilean transformation and Newton rather than with Maxwell's equations. We turn now to Einstien's first postulate and some consequences.

Four-space is right for formulations of special relativity

If the Galilean transformation is to be abandoned, what is to replace it? Einstein found the replacement must be the Lorentz transformation **L**. From the first postulate we have (in addition to the assumption that the relevant spaces are Euclidean) two fixed facts on which to base the transformation **L** that relates the fixed to the moving frame. We know that the origins of the frames move apart with velocity **u** (a velocity that will have the same value measured from either F or F'), and we know that light from the same flash would be measured to move with the same velocity in both frames. These anchors must be used to derive **L**. It will turn out that **L** is a transformation represented by an unitary matrix **L** that operates on vectors $\mathbf{r} = [x, y, z, ict]$ in a 4-space; thus changing from one frame of reference F to another F' is just a rotation in this 4-space with the amount of rotation determined by the relative velocity **u** with which the frames move apart. To find **L**, two steps are needed. The first step establishes that Einstein's first postulate implies that lengths in the 4-space just defined are preserved in going from F to F'. This in turn means that the **L** is a unitary transformation in this space because these are the only transformations that always preserve length. For the second step, an explicit form for the unitary transformation is found.

Picture a fixed frame of reference and a second frame F' sliding along F's x axis with velocity u (F''s y and z axes aligned with F's). Further, suppose that identical clocks in both frames are started at the instant at which the origins of F and F' coincide. This arrangement dictates that $y' = y$ and $z' = z$. Of course, the frames in general must be able to move in any direction relative to one another, but this convention on alignment simplifies the calculations and will not (except were noted) actually restrict the arguments. Now imagine that at $t = t' = 0$, a light is flashed at the origin. The light moves spherically in both F and F' with a constant velocity c (Einstein's first postulate), and a radius vector to the wave front satisfies the equation

$$x^2 + y^2 + z^2 = c^2t^2$$

in F and

$$x'^2 + y'^2 + z'^2 = c^2t'^2$$

in F'. These equations are conveniently written, in terms of the vector $\mathbf{r} = [x, y, z, ict]$ in F, as

$$\mathbf{r} \cdot \mathbf{r} = x^2 + y^2 + z^2 - c^2t^2 = 0$$

and, in terms of the corresponding vector in F', as

$$\mathbf{r}' \cdot \mathbf{r}' = 0.$$

The fourth component ict of the vector $\mathbf{r}$ is specifically chosen so that $(ict)^2 = -c^2t^2$ in order to make $\mathbf{r} \cdot \mathbf{r} = 0$. These 4-vectors reside in what is known as *Minkowski space* and turn out to be the appropriate way to formulate special relativity simply.

Lengths are preserved across space-time frames

The goal now is to demonstrate that Einstein's first postulate requires that lengths of 4-space vectors are preserved when one goes between frames F and F', although individual components are not, of course. The development that follows does not make explicit use of tensors, but the reader may wish to skim the section on page 192 to be aware of the tensors that are hidden from view. Length Δs in F is defined as

$$\Delta s^2 \equiv \Delta\mathbf{r} \cdot \Delta\mathbf{r} = \Delta x^2 + \Delta y^2 + \Delta z^2 - c^2\Delta t^2,$$

with a corresponding expression for F'. Preservation of length means that, for any F and F', lengths are invariant:

INVARIANCE OF LENGTH

$$\Delta s^2 = \Delta s'^2$$

This relationship will turn out to be central to relativity arguments. The job now is to show how it arises from Einstein's first postulate.

If F and F' define two Minkowski spaces, lengths in these spaces must be proportional to one another:

$$\Delta s = \lambda \Delta s'$$

We must find the value of λ. This can be done for any particular situation, and we choose the following simple one. In F a light pulse is shown at an angle from a point on the x axis onto a mirror situated l units above the axis with its reflective surface parallel to the x axis, so that the light is reflected back to the x axis. We measure Δx, the displacement along the x axis from the emission to receiving point, and Δt, the transit time for the light pulse. The F' frame is moving, as described above, along the x axis of the F frame (with $y = y'$ and $z = z'$), so $\Delta x'$ will be smaller than Δx; this is because F' is moving in the direction of travel of the light pulse and F' will have moved during the transit time of the light pulse. We also measured $\Delta t'$. For the F frame, the Minkowski space length is

$$\Delta s^2 = \Delta x^2 - c^2 \Delta t^2,$$

(remember that $\Delta y = \Delta z = 0$ because of the way the frames are aligned), and the same (primed) relation holds for the F' frame. The idea now is to calculate both lengths to determine the proportionality constant λ. The light must travel distance b in F, given by

$$b = 2\sqrt{l^2 + (\Delta x/2)^2},$$

so

$$\Delta x^2 = b^2 - 4l^2.$$

The speed of light is

$$c = \frac{b}{\Delta T},$$

and this equation can be used to eliminate b from the equation for Δx^2 two equations back to get

$$\Delta x^2 = c^2 \Delta t^2 - 4l^2.$$

Finally, this last relation permits us to eliminate Δx from the expression for Δs^2 to get the result that

$$\Delta s^2 = c^2 \Delta t^2 - 4l^2 - c^2 \Delta t^2 = -4l^2.$$

Precisely the same computations can be carried out in F', the only change being that all variables are primed; note, however, that c is the same in both frames, as is l. In F', then, $\Delta s'^2 = -4l^2$, so $\lambda = 1$. Thus the length of space-time vectors is preserved in going between the two 4-spaces defined by F and F' because $\Delta s^2 = \Delta s'^2$. Note that the constant relative velocity of the coordinate frames was not explicitly used and u appears nowhere in the equations. This assumption of constant velocity is necessary to have linear relations between the two coordinate frames. It is vital that l be the same in both frames; l does in fact have the same value, because this length is defined to be the same when the two frames are in register at time 0 and then cannot change because the motion is such to keep the axes all parallel.

The Lorentz transformation is the key

The next step is to get an explicit form for the Lorentz transformation for the special case of our coordinate systems being aligned as described above (the x' axis of F' slides along the x axis of F). This transformation, represented by a matrix, must be unitary in Minkowski space: only unitary transformations (with no stretching) preserve the length of all vectors. The matrix $\mathbf{L}$ thus must be unitary with a determinant equal to 1 and an adjoint equal to its inverse (see page 11). For this system, we know already that $y = y'$ and $z = z'$, so in the transformation matrix (which is, remember, unitary) the corresponding diagonal elements must be 1 and their off-diagonal elements 0. To see this, the thing to do is to write out the transformed vectors explicitly. For notational convenience, call $\mathbf{r} = [x_1, x_2, x_3, x_4] = [x, y, z, ict]$. Then a representative component of $\mathbf{L} = [a_{ij}]$ is

$$x'_i = \sum_j a_{ij} x_j.$$

When we explicitly examine these components one by one (remembering that $y' = y$, etc.), the transformation matrix **L** works out to be

$$\mathbf{L} = \begin{bmatrix} a_{11} & 0 & 0 & a_{14} \\ 0 & 1 & 0 & 0 \\ 0 & 0 & 1 & 0 \\ a_{41} & 0 & 0 & a_{44} \end{bmatrix}.$$

Specifically,

$$\begin{bmatrix} a_{11} & 0 & 0 & a_{14} \\ 0 & 1 & 0 & 0 \\ 0 & 0 & 1 & 0 \\ a_{41} & 0 & 0 & a_{44} \end{bmatrix} \begin{bmatrix} x \\ y \\ z \\ ict \end{bmatrix} = \begin{bmatrix} a_{11}x + a_{14}ict \\ y \\ z \\ a_{41}x + a_{44}ict \end{bmatrix} = \begin{bmatrix} x' \\ y' \\ z' \\ ict' \end{bmatrix}.$$

We now need to find explicit values for the various undetermined a_{ij} that appear in **L**. The approach used here has something of the character of "whatever works" to it, but the algebra is less complicated than that for some alternatives, and everything is done in an elementary way.

For $x_1 = x$ and $x_4 = ict$, transforming from F to F' with **L** gives

$$x' = a_{11}x + a_{14}ict,$$

and, for the inverse transform,

$$x = a_{11}x' + a_{41}ict'$$

(the unitary nature of **L** is used here). An observer sitting on the origin of F' will always be at $x' = 0$, and his x position, viewed from the origin of F, will simultaneously be $x = ut$, because F' is moving away from F with the constant velocity u. Substitute these values into the equation for x' above to get

$$0 = a_{11}ut + a_{14}ict,$$

so

$$a_{14} = -\frac{a_{11}u}{ic} \equiv i\gamma\frac{u}{c}.$$

The equation for x' can be written

$$x' = \gamma(x - ut),$$

where a_{11}, according to a usual convention, is defined thus:

$$a_{11} \equiv \gamma$$

The role of F and F' can be reversed—no coordinate system is preferred—so from F''s point of view, F is moving in a negative direction with a relative velocity u:

$$x = \gamma(x' + ut')$$

This means that

$$a_{41} = \frac{a_{11}u}{ic} = -a_{14} = -i\gamma\frac{u}{c}.$$

The next goal is to determine the explicit form of γ. Above the $x \to x'$ and $x' \to x$ transformations contain γ, but these equations also contain x, x', t and t'. We need more equations to eliminate two of these variables (so we can end up with two equations and two unknowns). Additional constraints come from noting that the x and x' positions of the light-flash wave front in F and F' are given by

$$x = ct \quad \text{and} \quad x' = ct'.$$

Use these to eliminate x and x' from the $F \leftrightarrow F'$ equations for the location of the wave front:

$$ct' = \gamma\overbrace{(ct - ut)}^{x'} = \gamma(c - u)t$$

$$ct = \gamma\overbrace{(ct' + ut')}^{x} = \gamma(c + u)t'$$

Divide through by c:

$$t' = \gamma(1 - \beta)t$$

$$t = \gamma(1 + \beta)t',$$

where $\beta \equiv u/c$. And eliminate t between the two equations to get

$$t' = \gamma^2(1 - \beta^2)t'.$$

The t's cancel, so this last equation gives an expression for γ:

$$\gamma = \frac{1}{\sqrt{1 - \beta^2}}$$

Finally we need to find a_{44}. Start with the $x \to x'$ equation,

$$x' = \gamma(x - ut),$$

and substitute $x = ct$ and $x' = ct'$ to get, for the special case of the position of the wave front,

$$ct' = \gamma(ct - ut).$$

Then divide by c and use the definition $\beta \equiv u/c$ to get

$$t' = \gamma(1 - \beta)t$$

for the $t \to t'$ transformation. Now, from the $x_4' = ict'$ component of **Lr**, one can get an expression for the same transformation (look back to see what a_{14} equals):

$$ict' = -i\gamma\beta x + a_{44}ict$$

At the wave front (because $x = ct$ there), after dividing by ic, this becomes

$$t' = \gamma\left(\frac{a_{44}}{\gamma} - \beta\right)t.$$

This means that $a_{44} = \gamma$, as can be seen by looking back at the earlier expression for $t' = \gamma(1 - \beta)t$. Thus, after dividing through by ic in the equation two equations back, the $t \to t'$ transformation becomes

$$t' = \gamma\left(t - \frac{\beta x}{c}\right).$$

We have treated, of course, only the special case of F' sliding along F's x axis, but one coordinate system could in general move in any direction with respect to another. If the moving frame is in the direction of one of the axes of F, the transformation is called *pure* or a *boost*. More general transformations are too messy to worry about.

Here is a summary of all the results on the Lorentz transformation:

KEY SPECIAL-RELATIVITY DEFINITIONS

$$\beta \equiv \frac{u}{c}$$

$$\gamma \equiv \frac{1}{\sqrt{1 - \beta^2}}$$

LORENTZ TRANSFORMATION
(BOOST)

$$x' = \gamma(x - ut)$$

$$t' = \gamma(t - \beta x/c)$$

$$x = \gamma(x' + ut')$$

$$t = \gamma(t' + \beta x'/c)$$

MINKOWSKI-SPACE ROW VECTOR

$$\mathbf{r} = [x, y, z, ict] = [x_1, x_2, x_3, x_4]$$

PROPER TIME

$$d\tau = \sqrt{1 - \beta^2}\, dt$$

(*Proper time* is defined on page 185.)

LORENTZ TRANSFORMATION MATRIX
(x BOOST)

$$\mathbf{L} = \begin{bmatrix} \gamma & 0 & 0 & +i\beta\gamma \\ 0 & 1 & 0 & 0 \\ 0 & 0 & 1 & 0 \\ -i\beta\gamma & 0 & 0 & \gamma \end{bmatrix}$$

Note that $\mathbf{L}$ depends continuously on β, the relatively velocity of the two reference frames. The Lorentz transformations thus form a continuous group known as the *Lorentz group* of transformations. These are a subgroup of a larger group, the *Poincaré* transformations. Poincaré transformations involve a rotation (like the Lorentz transformations) and also a translation, and they too preserve the velocity of light. The special class of transformations we have treated, those where the two frames have parallel axes, are termed *pure* Lorentz transformations or *boosts*, as has been noted above. In the limit of low relative velocities ($\beta \to 0$), the Lorentz transformation approaches the Galilean transformation, so the distinction between relativistic and classical physics vanishes.

The Lorentz transformation can lead to strange results

Einstein's first postulate and the resulting Lorentz transformations have some very peculiar consequences, two of which are the *Fitzgerald contraction* and *time dilation.* These are considered in turn.

Suppose that a rod of length $l = \Delta x$ in F is parallel to the x axis (the direction of F' movement) and this rod is measured in F and in F'. According to the LORENTZ TRANSFORMATION,

$$x' = \gamma(x - ut),$$

so a length difference, measured at $t = t' = 0$, is described by the *Fitzgerald contraction*:

FITZGERALD CONTRACTION

$$\Delta x' = \gamma \Delta x$$

If the rod has length $l' = \Delta x'$ in the moving frame F', then the length $l = \Delta x$, measured in F, would be

$$l = \frac{l'}{\gamma} = l'\sqrt{1 - \beta^2}.$$

The moving rod, when measured in F, has a length contracted by a factor that approaches 0 as u approaches c in comparison with its length measured in F'. Thus moving objects are found to be shorter in the stationary frame than they are in the moving frame.

Suppose that a light is flashed twice at the origin of F with the interval between flashes Δt. The time between these flashes $\Delta t'$ is measured at the origin of F'. Since the LORENTZ TRANSFORMATION specifies that

$$t' = \gamma(t - \beta x/c),$$

at a fixed x location we have the equation for *time dilation*:

TIME DILATION

$$\Delta t' = \gamma \Delta t$$

Because $\gamma \geq 1$, the time between the same two events (the interval between light flashes) is longer in the moving frame than in the fixed frame. Thus, moving clocks tick more slowly, and moving objects are longer in the moving frame.

Because time varies between fixed and moving frames, a definition of time that is the same for all reference frames is useful. The timelike quantity defined in this way is called *proper time*, and a proper-time interval is denoted by $d\tau$. Suppose you are moving in frame F' with velocity u. According to the time-dilation relation above, a brief interval dt' would be longer for you by a factor γ than the corresponding dt measured in a resting frame F. If a *proper* time interval is defined as

$$d\tau \equiv \frac{dt'}{\gamma} = dt'\sqrt{1 - \beta^2} = dt'\sqrt{1 - u^2/c^2},$$

and if this time is used by observers in all frames, then they will all measure time intervals that are numerically the same, irrespective of the reference frame in which the time measurement is made. Proper time will be important later in deciding how to treat classical mechanics in the relativistic framework.

Theories Must Be Covariant

As noted at the outset, Newtonian physics is covariant (form invariant) under Galilean transformations, but electromagnetism is not. One theory had to be wrong, and Einstein decided to abandon Newtonian physics and the principle that laws of physics are covariant under the Galilean transformation. The preceding section presented Einstein's alternative, special relativity. Now the development focuses on Einstein's second postulate—physical laws should have the same form in all inertial frames—and proceeds in two ways. The first is to reformulate Newtonian mechanics to conform to special relativity. The second is to cast Maxwell's equations in a form that is, in the terminology generally used, *manifestly covariant*. The latter means using a tensor formulation of electricity and magnetism, where it can be seen by inspection that the form of relations is preserved under Lorentz transformations.

The reformulation of Newtonian mechanics

The problem is to find a covariant (that is, Lorentz-covariant) formulation of mechanics that reduces to the classical equations as velocities become sufficiently small. Throughout, attention will be restricted to the special Lorentz transformation, known as a boost, that was used in the preceding section; that is, the F and F' frames are aligned so as to have parallel axes and F' slides along F's x axis with a constant velocity u.

We seek the appropriate generalization of NEWTON'S SECOND LAW, $f = ma = \dot{p}$. The generalization of the momentum p turns out to be fairly easy and natural, but the covariant formulation of the force, f, is harder. The difficulty here really has to do with Newton, not with relativity: each force has to be dealt with separately, since no comprehensive classical theory of the forces that act on mechanical systems is available. Electromagnetic forces are considered later (their covariant formulation is immediate), but for the time being, we just assume that the forces in Newton's law are no problem for special relativity. The starting place for the generalization of mechanics, then, is to identify relativistic momentum.

What is needed is a covariant formulation of momentum that reduces to Newton's momentum in the limit where particles have small velocity. Start with the vector in Minkowski space $\mathbf{r} = [x, y, z, ict]$ and its differential $d\mathbf{r} = [dx, dy, dz, ic\,dt]$, both of which are covariant under Lorentz transformation because $\mathbf{L}$ preserves lengths in 4-space. The infinitesimal proper-time interval $d\tau$ has precisely the same value in every inertial frame and additionally reduces to dt when a particle has a small velocity, so the derivative of $\mathbf{r}$ with respect to proper time is a promising approach.

Proper time, however, needs some further refinement before it can be used to extend the definition of momentum. Because of the time dilation described earlier, time goes slower in a moving frame than it does in one at rest. Although this was not known at the time Einstein developed the theory, recall that rapidly moving unstable particles actually decay at a slower rate (as measured in the nonmoving laboratory) than slowly moving ones: these relativistic effects on time are real and not merely conventions of the formalism. Thus the time associated with a particle needs to be determined by a clock that is moving along with the particle if we are to get the physics right. Start by imagining a particle moving along the x axis in the laboratory with a constant velocity $\dot{x}$. We can just as well think of the particle as fixed at the origin of a frame that is moving, relative to the laboratory, with velocity $\dot{x}$. In this way, the clock associated with the moving frame, the clock that reads the proper time t', will be attached to the particle.

How can someone in the laboratory read the time on the particle's clock? The particle's time, that is, the time associated with the moving coordinate system in which the particle is at rest, is t'. Also, the particle's 4-space position is always $\mathbf{r}' = [0, 0, 0, ict']$, because, by construction, the particle never leaves the origin of its moving coordinate system. Now the proper time $d\tau^2 = -\mathbf{r}' \cdot \mathbf{r}'/c^2 = dt'^2$ (the spatial coordinates do not appear, because the particle is constantly at the origin of its moving frame) has the same value in all Lorentz frames, so an observer can simply measure $d\tau$ in the laboratory frame to read the particle's clock. The equation that relates time intervals in the laboratory to $d\tau$, and thus directly to dt', which is the same as the proper time in this situation, is

$$dt' = d\tau = dt\sqrt{1 - \dot{x}^2/c^2} = dt\sqrt{1 - \beta^2}$$

(see the discussion of proper time in the preceding section). Here $\dot{x}$ is the particle's velocity measured in the laboratory (with the laboratory clock) and $\beta = \dot{x}/c$.

The natural definition for the x component of relativistic momentum, then, would be (and is) $\pi = m(dx/d\tau)$ (easily extended at once to other components). This identification of momentum uses the particle's clock (which it should, because we know from particle-decay experiments that time proceeds at a different rate for particles in rapid motion). This relativistic momentum is covariant under the Lorentz transformation (π is a covariant position vector divided by a scalar with the same value in all coordinate systems), and it reduces to Newton's momentum when $\dot{x}$ is small compared to c (because $dt' \to dt$). Note than β here is the particle's velocity, measured in units of c.

Here is the tricky part: A particle might well move with a time-varying velocity rather than the constant velocity assumed in the previous discussion, so its clock would not be situated in an inertial frame. Thus the use of the particle's dt' to define momentum for the general situation does give a mathematically valid covariant formulation—momentum defined in this way will have the same form in all coordinate systems for which the speed of light is constant—but a new physical assumption is involved because we are using time defined by accelerating clocks, and this is not included in Einstein's original postulates. This new assumption of the validity of time measured by accelerated clocks that ride along with accelerated particles is needed to give a covariant mechanics. Note that in this context β, the particle's velocity measured in units of c, has a somewhat different meaning than it did previously. Before β was a constant used in the Lorentz transformation to go from a fixed to a moving frame of reference, but now β is a possibly variable velocity of a particle moving in the laboratory frame. With several particles present, each would, of course, have its own β. This formulation has been abundantly confirmed. We thus identify the *relativistic momentum*:

RELATIVISTIC MOMENTUM

$$\pi = \frac{dm\mathbf{r}}{d\tau} = \frac{m}{\sqrt{1 - \dot{r}^2/c^2}} \frac{d\mathbf{r}}{dt}$$

Here $\mathbf{r}$ is a position vector in Minkowski space, $\dot{r}$ is the particle's velocity measured in the laboratory, and m is the rest mass of the particle. Thus the Minkowski space vector representing momentum is, in components,

$$\pi = \left[\frac{dmx}{d\tau}, \frac{dmy}{d\tau}, \frac{dmz}{d\tau}, icm\frac{dt}{d\tau}\right]$$

or, with velocities as measured in the (nonmoving) laboratory ($\dot{x} = dx/dt$, etc.),

$$\boldsymbol{\pi} = [m'\dot{x}, m'\dot{y}, m'\dot{z}, icm'],$$

with

$$m' = \frac{m}{\sqrt{1 - \dot{r}^2/c^2}}.$$

Having identified an appropriate relativistic momentum, we now need to turn to a covariant formulation of mechanics and examine the conservation of momentum and of kinetic energy.

As noted earlier, the problem in trying to find a covariant formulation of Newton's second law is with the forces. Later we will see that electromagnetic forces can be described appropriately, and if we can then pit this electromagnetic force against other, as yet unidentified, forces to achieve an equilibrium, we will have demonstrated that other forces must also have a covariant formulation. For the time being, however, we just assume that Lorentz-invariant forces can be identified. The required covariant formulation, then, is the *relativistic version of Newton's second law*:

RELATIVISTIC VERSION OF
NEWTON'S SECOND LAW

$$\frac{d\boldsymbol{\pi}}{d\tau} = \boldsymbol{\phi}$$

Here ϕ is called the *Minkowski* force; it must, of course, reduce to a classical force in the limit of low velocities. Here and in the remainder of this subsection, I treat π and ϕ sometimes as the x component of a 4-vector and other times as the vectors themselves; in this later case, the vector will be boldfaced, and the components will be π_j and ϕ_j, as usual.

We need to translate the preceding equation to laboratory coordinates. Remember that the particle's time $d\tau$ is related to laboratory time by the factor $\sqrt{1 - \dot{x}^2/c^2}$ ($\dot{x}$ is the particle's velocity measured in laboratory space-time), so the relativistic version of Newton's second law becomes

$$\frac{m}{\sqrt{1 - \dot{x}^2/c^2}} \frac{d\dot{x}}{dt} = \phi\sqrt{1 - \dot{x}^2/c^2}$$

(the term $\sqrt{1 - \dot{x}^2/c^2}$ on the right arises when $d\tau \to dt$). The relativistic force is thus $\phi\sqrt{1 - \dot{x}^2/c^2}$. Note that in the absence of forces, $d\pi/dt = 0$, so momentum is conserved. This is also true if a number of particles are present.

How does momentum act under Lorentz transformations? (Now I treat $\boldsymbol{\pi}$ as a vector.) The dot product $\boldsymbol{\pi} \cdot \boldsymbol{\pi}$ is a scalar and has the same value in all frames, because scalars are unchanged by rotations. Explicitly ($\boldsymbol{\pi}$ is a vector, but assume, for convenience, that the particle moves along the x axis),

$$\begin{aligned}\boldsymbol{\pi} \cdot \boldsymbol{\pi} &= \frac{m^2\dot{x}^2}{1 - \dot{x}^2/c^2} - \frac{m^2c^2}{1 - \dot{x}^2/c^2} \\ &= \frac{m^2\dot{x}^2 - m^2c^2}{1 - \dot{x}^2/c^2} = \frac{m^2c^2(\dot{x}^2/c - 1)}{1 - \dot{x}^2/c^2} = -m^2c^2.\end{aligned}$$

Thus relativistic momentum is conserved in a particular frame over time and has a constant value for all frames. This value will later be related to energy.

Up to this point we have identified a relativistic momentum, have given a covariant formulation of mechanics (pending the proper identification of forces), and have shown (again, provisionally, since this depends on the relativistic second law) that momentum is conserved and invariant under Lorentz transformations. Now we turn to the relativistic formulation of the kinetic energy of a particle.

Classically, kinetic energy T obeys the equation

$$\frac{dT}{dt} = \frac{d}{dt}\left(\frac{m\mathbf{v}\cdot\mathbf{v}}{2}\right) = m\dot{\mathbf{v}}\cdot\mathbf{v} = \mathbf{f}\cdot\mathbf{v},$$

where $\mathbf{f}$ is, according to Newton, the force and $\mathbf{v}$ the particle's velocity. Identify T by writing out the relativistic version of this equation. Instead of $\mathbf{v}$, use the velocity $\boldsymbol{\pi}/m$ (still a vector), and instead of $\mathbf{f}$, use $d\boldsymbol{\pi}/d\tau$ to write the relativistic version of $\mathbf{f}\cdot\mathbf{v}$:

$$\mathbf{v}\cdot\mathbf{f} \rightarrow \underbrace{\frac{\boldsymbol{\pi}}{m}\cdot\frac{d\boldsymbol{\pi}}{d\tau}}_{\frac{1}{2m}\frac{d}{dt}(\boldsymbol{\pi}\cdot\boldsymbol{\pi})} = \frac{\boldsymbol{\pi}}{m}\cdot\boldsymbol{\phi}$$

The left side of this equation equals

$$\frac{1}{2m}\frac{d}{d\tau}(\boldsymbol{\pi}, \boldsymbol{\pi}),$$

and this vanishes because $\boldsymbol{\pi}\cdot\boldsymbol{\pi}$ is the constant $-m^2c^2$ (that is, a scalar whose value is unaffected by rotations in Minkowski space). The right side of the equation is

$$\frac{\boldsymbol{\pi}}{m}\cdot\boldsymbol{\phi} = \frac{\mathbf{f}\cdot\mathbf{v}}{1 - v^2/c^2} + \frac{ic\phi_4}{\sqrt{1 - v^2/c^2}}$$

(write out the dot product in terms of components). This can be solved for the fourth component of the Minkowski force to get

$$\phi_4 = \frac{i\mathbf{f}\cdot\mathbf{v}}{c\sqrt{1 - v^2/c^2}}.$$

Another expression for the fourth component of the Minkowski force comes from the relativistic Newton's second law. The fourth component of $d\boldsymbol{\pi}/d\tau$ satisfies the equation

$$\left(\frac{d\boldsymbol{\pi}}{d\tau}\right)_4 \rightarrow \frac{d}{dt}\left(\frac{icm}{\sqrt{1 - v^2/c^2}}\right) = \phi_4\sqrt{1 - v^2/c^2}$$

(note the change on the left to laboratory time, which gives $\sqrt{1 - v^2/c^2}$ on the right). Eliminate ϕ_4 between these two equations to get

$$\frac{d}{dt}\left(\frac{mc^2}{\sqrt{1 - v^2/c^2}}\right) = \mathbf{f} \cdot \mathbf{v}.$$

Thus we can recognize the relativistic kinetic energy as

$$T = \frac{mc^2}{\sqrt{1 - v^2/c^2}}.$$

This expression does not immediately look anything like the usual kinetic energy, but it should reduce to the classical kinetic energy in the limit of a small v. So, expand T to first order:

$$T \approx mc^2 + \frac{mv^2}{2}$$

The second term is indeed the classical kinetic energy, and the first term is just a constant added to all kinetic energies.

One might be tempted to simply ignore the constant (energy is only determined to an additive constant anyway), but Einstein realized that it could be interpreted as the rest energy $\mathscr{E}$ associated with a mass m; he expressed this relation in his most famous equation:

EINSTEIN'S
MOST FAMOUS EQUATION

$$\mathscr{E} = mc^2$$

Note that the "silent" constant $\mathscr{E}$ is to added to all of the classically calculated values of energy E. Particle experiments have shown that this is valid. Look back and note that the relativistic kinetic energy T is related to the forth component of the relativistic momentum $\boldsymbol{\pi}$:

$$\pi_4 = icm' = \frac{icm}{\sqrt{1 - v^2/c^2}} = \frac{iT}{c}$$

For a free particle, the entire energy is kinetic, so $E = T$. Thus $c^2\pi_4^2 = -E^2$ (according to the equation above), and $\boldsymbol{\pi}^2 = -m^2c^2$ (according to the relation on page 189). This means that

$$c^2\boldsymbol{\pi} \cdot \boldsymbol{\pi} = c^2\mathbf{p} \cdot \mathbf{p} + c^2\pi_4^2 = c^2\boldsymbol{\pi}^2 - E^2 = -c^4m^2,$$

so

$$E^2 = c^2p^2 + c^4m^2,$$

where $\mathbf{p} = m'\mathbf{r}$ for the relativistic mass m' and the 3-space position vector $\mathbf{r}$. If the mass happens to be 0, as it is for the photon and some other particles, the momentum is $p = E/c$. Compare this to the classical momentum $p = \sqrt{2mE}$ for massive particles.

Why use tensors?

Einstein's second postulate is that physical laws must have the same form in every inertial frame. What we need, then, is a way of formulating these laws that makes it immediately apparent whether a particular relationship is or is not consistent with relativity theory. Tensors have just the characteristic that they are unchanged in form when the coordinate system is rotated, so tensors are the appropriate quantities for formulating theories to be tested for consistency with Einstein's second postulate. Thus tensors (see page 53) are used in relativity theory to make immediately clear whether a particular theory is covariant or not. What this means is explained in this section.

I actually employed this approach in the preceding development but did not make the use of tensors explicit; the relativistic momentum was specifically constructed to be a tensor. To see why one requires a tensorial formulation of physical laws in the context of special relativity, we reexamine the way relativistic mechanics was developed.

The key to the formulation of special relativity presented in the first section was the observation that the Lorentz transformations required to insure that light has the same speed in all inertial frames are just rotations in Minkowski 4-space. In the treatment of mechanics, we used this fact by finding expressions for momentum and force that, by virtue of their Minkowski 4-vector nature, we knew would be covariant under Lorentz transformations. Remember: all scalars and all legitimate vectors in a vector space (here, Minkowski space) are tensors for that space. Thus the transformation used to change frames of reference has no effect on the form of the physical law, and Einstein's second postulate is satisfied (physical laws look the same in all reference frames). The test for the covariance of a theory, then, is that its form be unchanged by 4-space rotations. Any formulation that, under Lorentz transformations, behaves like 4-space vectors will be clearly covariant: this is just the definition of tensors (see page 54).

To produce a relativistic version of classical mechanics, we developed 4-space vectors that reduced to the classical expressions for low particle velocities, but these vectors, by virtue of the way they were constructed, transformed just like the vectors in Minkowski space. Not every four-component vector has this property; that is, not every object with four components is a legitimate vector in Minkowski space. An example of a four-component object that does not reside in Minkowski space is $d\mathbf{r}/dt$, where $\mathbf{r}$ is a position vector in Minkowski space. Although $d\mathbf{r}$ transforms like $\mathbf{r}$, and thus would be covariant, t is one of the vector components (for the rest frame $r_4 = ict$), so dividing spatial components by dt gives a four-component object that, under a Lorentz transformation, no longer behaves like $\mathbf{r}$ itself; $d\mathbf{r}/dt$ is not covariant. Scalars and vectors in Minkowski space are tensors of rank 0 and 1, but we did not have to explicitly recognize their tensorial character in the preceding discussion because mechanics is formulated in terms of vectors in any case. Electricity and magnetism requires second- and forth-rank tensors (in addition to scalars and vectors), so when casting electromagnetism in a manifestly covariant form, our next task, one cannot avoid the explicit use of tensors.

Before continuing, however, it might be helpful to make some comments about the word *covariant*. This term arises from the need to distinguish between the behavior of scalars, which are invariant under rotations (a scalar has precisely the same value, no matter how its coordinate system is oriented), and that of vectors, whose components change when their coordinate system is rotated, even though the vector properties (length and direction relative to other vectors) are unchanged. Because they are not truly invariant, such unchanged vectors are termed "co-variant" to indicate that they represent the same "thing" before and after their coordinate system is rotated, despite the fact that the magnitude of their components can individually can change. We then speak of a "covariant formulation" of, say, mechanics, but strictly speaking, we should say that the theory is covariant under the Lorentz transformation, for that is what is intended. The property of covariance is relative to the transformation being used and the vector space in which the vector resides (note that we earlier used the notion of covariance under the Galilean transformation). To make things still a little more complicated, not only is "covariant" used in a sometimes imprecise and confusing way, but the word even has different meanings in different contexts. "Covariant" is, for example, used to refer to a type of tensor (in contradistinction to "contra-variant" tensors), but this meaning is unrelated to the one we have been using.

The covariant form of electricity and magnetism*

Our next goals are to show that equations for the electromagnetic field are covariant, to demonstrate that the Lorentz force on a moving charged particle is covariant (remember that this was needed to complete the relativistic extension of Newton's second law), and to cast Maxwell's equations in a manifestly covariant form. The first step is to identify the relevant Minkowski-space vectors for the electromagnetic field. This we do by writing a tensor version of the equations for the vector potentials, which enables us to see that Maxwell's theory has to be a covariant one. Next we show that the force on a moving charge is covariant. Finally, we put Maxwell's equations into a manifestly covariant, or tensorial, form.

Recall (see page 95) that Maxwell's equations boil down to, for the Lorentz gauge, the four equations for the potentials: three equations for **A**,

$$\nabla^2 \mathbf{A} - \frac{1}{c^2}\partial_t^2 \mathbf{A} = -\frac{4\pi}{c}\mathbf{J},$$

and one equation for V,

$$\nabla^2 V - \frac{1}{c^2}\partial_t^2 V = -4\pi\rho.$$

We begin by demonstrating that these equations are covariant. The idea is to write what appear to be tensor versions of the equations and then to demonstrate that what look like tensors actually are.

Start with the differential operator in the preceding equations:

$$\nabla^2 - \frac{1}{c^2}\partial_t^2 \equiv \Box\cdot\Box = \Box^2 \equiv \sum_\mu \partial_\mu^2$$

Here $\Box$ is the 4-space gradient operator

$$\Box = \left[\partial_x, \partial_y, \partial_z, \frac{1}{ic}\partial_t\right].$$

The operator $\Box^2$ transforms like a scalar in Minkowski space because $\Box$ is defined to have components of a Minkowski-space vector; that is, $\Box^2$ acts like a tensor. Two conventions are used in this section: (1) Greek subscripts take on the values $1,\ldots,4$, and (2) $\partial_\mu \equiv \partial/\partial x_\mu$ (note that $\partial_4 = (1/ic)\partial_t$). Using this notation and the 4-vector $\mathscr{A}$ (which we do not yet know to be covariant)

$$\mathscr{A} \equiv [A_\nu] \equiv (A_x, A_y, A_z, iV],$$

we can write the left side of the four equations for **A** and V,

$$\nabla^2 \mathbf{A} - \frac{1}{c^2} \partial_t^2 \mathbf{A} = -\frac{4\pi}{c} \mathbf{J}$$

and

$$\nabla^2 V - \frac{1}{c^2} \partial_t^2 V = -4\pi\rho,$$

as $\square^2 A_\nu$. Now consider the right side of the equations for **A** and V. If we define the generalized current, another 4-vector (not yet shown to be covariant), as

$$\mathbf{j} = \frac{4\pi}{c} [J_x, J_y, J_z, ic\rho],$$

then the equations for potentials become

$$\square^2 A_\nu = -j_\nu.$$

Further, the defining equation for the Lorentz gauge, $\nabla \cdot \mathbf{A} + (1/c)\partial_t V = 0$, can be expressed as

$$\square \cdot \mathscr{A} \equiv \partial_\mu A_\mu = 0.$$

(Note that the Einstein summation convention, defined on page 55, is used.) These look as if they might be tensor equations, but their tensorial nature must be established. Since $\sum \partial_\mu^2$ does behave like a tensor in Minkowski space, we need only establish that the 4-vector **j** does also; if so, then A_ν will be a tensor (see the section "Cartesian Tensors," page 53).

The 4-vector **j** is the product of the constant $4\pi/c$ and $\rho[v_x, v_y, v_z, ic]$. Start with $d\mathbf{r}/dt$ ($\mathbf{r} = [x, y, z, ict]$; assume, for simplicity, a boost along the x axis with relative velocity β), and define the vector **h** as

$$\mathbf{h} \equiv [v_x, v_y, v_z, ic] = \frac{d\mathbf{r}}{dt} = \frac{1}{\sqrt{1-\beta^2}} \frac{d\mathbf{r}}{d\tau}.$$

Remember that $d\mathbf{r}/d\tau$ is covariant, so $\mathbf{h} = d\mathbf{r}/dt$ is not because **h**'s value varies with velocity β. Thus **j** is the product of a constant and the two factors ρ and **h**:

$$\mathbf{j} = \frac{4\pi}{c}\rho\mathbf{h}$$

We now wish to demonstrate that, although neither $\mathbf{h}$ nor ρ is covariant, the product $\rho\mathbf{h}$ (and thus $\mathbf{j}$) is.

Consider the charge density ρ, the number of charges per unit volume. Experimental evidence supports the idea that charge is unaffected by a particle's velocity. For example, a hydrogen molecule has no net charge, yet its electrons are moving rapidly around an unmoving nucleus. If motion altered the magnitude of charges, the two moving electrons in H_2 should not exactly balance the two stationary nuclear charges. Now picture a quantity of charge dq contained inside the box $dxdydz$, so the charge dq within the box will be $d\rho dxdydz$. The effect of the boost on this tiny box containing the charge will be to change the box dimensions along the x axis and thus to alter the charge density $d\rho$. Both dq and (because of the direction of the motion) $dydz$ are unchanged in the moving frame, but $dx = dx'\sqrt{1-\beta^2}$. This means that the charge density changes with motion from ρ to $\rho' = \rho\sqrt{1-\beta^2}$. The vector $\rho\mathbf{h}$ in the moving frame is thus $\rho'(d\mathbf{r}'/dt) = \rho(d\mathbf{r}/d\tau)$, which is covariant because the $\sqrt{1-\beta^2}$ in ρ' cancels the $1/\sqrt{1-\beta^2}$ in $\mathbf{h}$ and the product is unaffected by the movement. Thus $\mathbf{j}$, a constant times $\mathbf{h}$, is a tensor. Since $\Box^2$ and $\mathbf{j}$ are both tensors, A_ν must be a first-rank tensor. Further, because the four equations for $\nabla^2\mathbf{A}$ and $\nabla^2 V$, together with the defining equation for the Lorentz gauge, are equivalent to Maxwell's equations, Maxwell's equations themselves must be covariant. Note that because Maxwell's equations are form-variant under Lorentz transformations, they cannot be form-invariant under the Galilean transformation, as discussed on page 175.

Next we turn to the force on a particle with charge q moving with velocity $\mathbf{v}$. We wish to establish that the Lorentz force is covariant. Classically, this force is

$$\mathbf{f} = \dot{\mathbf{p}} = q\left(\mathbf{E} + \frac{1}{c}\mathbf{v} \times \mathbf{B}\right)$$

(see the chapter "Electricity and Magnetism"). To find the relativistic version of this equation, we must replace the force $\mathbf{f}$ with the Minkowski force ϕ, the time derivative of momentum $\dot{\mathbf{p}}$ with the 4-vector $d\boldsymbol{\pi}/d\tau$, and the particle velocity $\mathbf{v}$ with its relativistic counterpart $\mathbf{u} = (d/d\tau)(x, y, z, ict)$. Start by looking at the fourth component of $d\boldsymbol{\pi}/d\tau$; the first three components come at once from the relativistic generalization of the Lorentz-force equation. As was noted earlier, the fourth component of $\boldsymbol{\pi}$ is proportional to the particle's relativistic kinetic energy T, $\pi_4 = iT/c$, so $d\boldsymbol{\pi}/d\tau$ is proportional to the time derivative of the work done on the particle by the field. This rate of change can be written, for work W and the 4-space position vector $\mathbf{r}$, as

$$\frac{dW}{d\tau} = q\mathbf{E}\cdot\frac{d\mathbf{r}}{d\tau} = q\mathbf{E}\cdot\mathbf{u},$$

with relativistic velocity $\mathbf{u} = d\mathbf{r}/d\tau$. Thus the fourth component of $d\boldsymbol{\pi}/d\tau$ is

$$\frac{d\pi_4}{d\tau} = \frac{iq}{c}\mathbf{E}\cdot\mathbf{u}.$$

The Lorentz-force equations for a particle with relativistic velocity $\mathbf{u}$, combined with this last equation for the fourth component of $d\boldsymbol{\pi}/d\tau$, can be expressed, component by component, as

$$\frac{d\pi_1}{d\tau} = \frac{q}{c}(0 + u_2B_3 - u_3B_2 - iu_4E_1) \quad \rightarrow \frac{q}{c}(E_x + (\mathbf{v}\times\mathbf{B})_x),$$

$$\frac{d\pi_2}{d\tau} = \frac{q}{c}(-u_1B_3 + 0 + u_3B_1 - iu_4E_2) \rightarrow \frac{q}{c}(E_y + (\mathbf{v}\times\mathbf{B})_y),$$

$$\frac{d\pi_3}{d\tau} = \frac{q}{c}(u_1B_2 - u_2B_1 + 0 - iu_4E_3) \quad \rightarrow \frac{q}{c}(E_z + (\mathbf{v}\times\mathbf{B})_z),$$

$$\frac{d\pi_4}{d\tau} = \frac{q}{c}(iu_1E_1 + iu_2E_2 + iu_3E_3 + 0) \rightarrow \frac{iq}{c}\mathbf{v}\cdot\mathbf{E}$$

(remember that $u_4 = ic(dt/d\tau)$). These equations reduce to the nonrelativistic versions of the Lorentz-force equation for low velocities, as indicated. The preceding equations for the Minkowski force ϕ (see the section on relativistic mechanics, p. 189) can be written compactly as the *relativistic Lorentz force on a moving charge*:

RELATIVISTIC LORENTZ FORCE
ON A MOVING CHARGE

$$\phi_\nu = \frac{d\pi_\nu}{d\tau} = \frac{q}{c}F_{\nu\mu}u_\nu$$

Here $F_{\nu\mu}$ is represented by the *field-strength tensor*:

FIELD-STRENGTH TENSOR

$$[F_{\nu\mu}] = \begin{bmatrix} 0 & -B_3 & B_2 & iE_1 \\ B_3 & 0 & -B_1 & iE_2 \\ -B_2 & B_1 & 0 & iE_3 \\ -iE_1 & -iE_2 & -iE_3 & 0 \end{bmatrix}$$

Because both $d\pi/d\tau$ and $\mathbf{u}$ are vectors in Minkowski space, $F_{\nu\mu}$ must be a second-rank antisymmetric ($F_{\nu\mu} = -F_{\mu\nu}$) tensor.

Finally, we can exhibit Maxwell's equations in a manifestly covariant form. The field-strength tensor may, as can be verified by writing out components, be used to compactly write two of MAXWELL'S EQUATIONS (Coulomb's law and Ampère's law plus the conservation of charges, for $\nabla \cdot \mathbf{E}$ and $\nabla \times \mathbf{B}$):

$$\partial_\mu F_{\nu\mu} = j_\nu$$

Written out in components, this last equation is

$$-\partial_y B_z + \partial_z B_y - \frac{1}{c}\partial_t E_x = J_x,$$

$$\partial_x B_z - \partial_z B_x - \frac{1}{c}\partial_t E_y = J_y,$$

$$-\partial_x B_y + \partial_y B_x - \frac{1}{c}\partial_t E_z = J_z,$$

$$-i\partial_x R_x - i\partial_y E_y - i\partial_z E_z = -i4\pi\rho,$$

which give just the two required Maxwell's equations.

The other two of Maxwell equations (Faraday's law, for $\nabla \times \mathbf{E}$, and no magnetic monopoles, for $\nabla \cdot \mathbf{B}$) are not as neat. They work out to be

$$\partial_\mu F_{\nu\theta} + \partial_\theta F_{\mu\nu} + \partial_\nu F_{\theta\mu} = 0,$$

where μ, ν, θ each can be 1, 2, 3, or 4 (but all are different). For example, for $(\mu, \nu, \theta) = (1, 2, 3)$, these last equations turn out to be

$$-\partial_x B_x - \partial_z B_z - \partial_y B_y = 0,$$

which is $\nabla \cdot \mathbf{B} = 0$. For $(\mu, \nu, \theta) = (1, 2, 4)$, they are

$$i\partial_x E_y - \frac{1}{ic}\partial_t B_z - i\partial_y E_x = 0,$$

which is the z component of $\nabla \times \mathbf{E} + (1/c)\partial_t \mathbf{B} = 0$. By writing out the rest of the components, one can verify that the tensor equations do indeed represent the remaining two of Maxwell's equations. Thus, in manifestly covariant form, we have the *tensor version of Maxwell's equations*:

TENSOR VERSION
OF MAXWELL'S EQUATIONS

$$\partial_\mu F_{\nu\mu} = j_\nu$$

$$\partial_\mu F_{\nu\theta} + \partial_\theta F_{\mu\nu} + \partial_\nu F_{\theta\mu} = 0$$

7 Quantum Field Theory

The goal of quantum field theory is to account for the properties of subatomic particles that make up matter and for the forces between these particles. This theory is a natural extension of quantum mechanics and, indeed, originally grew out of attempts to provide a quantum-mechanical description of the way charged particles like electrons interact with the electromagnetic field. This chapter presents the general structure of field theory.

To modify the quantum mechanics presented in chapter 4 so that it could account for the behavior of charged particles moving in an electromagnetic field, two main problems needed to be solved. First, the particles, like electrons, that were to be the subjects of this theory often had to be supposed to move at speeds approaching that of light, so some relativistic extension of quantum theory was required. SCHRÖDINGER'S EQUATION is not covariant, and how to develop the appropriate extension was not obvious. Second, the electromagnetic field is classical, and one does not immediately know how to include it in the quantum-mechanical picture. These difficulties, and some others that arose in the course of developing quantum electrodynamics, were surmounted and an enormously successful theory emerged. One of the successes of this theory was the fact that particles, light quanta, came out of the theory in a natural way as a consequence of a quantum treatment of the electromagnetic field. The obvious idea, then, was that all particles arise from their own field in the same way: photons from the electromagnetic field, electrons from an electron field, mesons from a meson field, etc. The working out of this idea has preoccupied physics for the past five decades.

One of the most serious problems that the theory must confront is that the number and nature of particles in the microphysical world is not fixed. Rather, subatomic particles are created and destroyed, and this fact must become central to any suitable theory of subatomic particles. That the numbers of particles can vary presents a severe problem for the quantum mechanics outlined in chapter 4. The reason is that the number of particles present determines the number of variables included in the wave function, and a new wave function would have to be calculated for every creation and destruction event. Traditional quantum mechanics is simply not suited for this situation.

So the immediate goals for the extension of quantum mechanics are (1) to include classical fields, like electromagnetic and (one hopes) gravitational fields, within the quantum-mechanical framework, (2) to generalize quantum mechanics to include the domain of relativity theory, and (3) to take account of the creation and annihilation of particles. The final objective is to devise an unified theory that will explain all of the forces of nature and the particles on which they act. This chapter surveys the methods currently used for achieving the first three goals and indicates the general approach to the main objectives of the unified theory.

The extension of quantum theory to include the electromagnetic field is conceptually easy and direct: the amplitude for a transition from one field configuration to another is found by using the KEY-PATH INTEGRAL with the classical Lagrangian for the electromagnetic field. Our approach is no different from what we have done earlier except that, instead of the classical Lagrangian that corresponds to Newton's equation, we use the Lagrangian that corresponds to Maxwell's equations. This method works. How, then, do we extend the theory to particles other than photons, say to mesons? The trick is to replace the Lagrangian for the electromagnetic field with one for the meson field. All of the physics, then, is in the selection of the Lagrangian to use. For this chapter we will simply assume that some method can give the Lagrangian we need for the field—and thus for the type of particle—and focus on the way one develops the theory starting from a Lagrangian.

Although I will talk about the electromagnetic field, the actual calculations will be for a simpler case. The problem is that the Lagrangian for the electromagnetic field is complicated, and so the fussy details involved in treating this field obscures the essential aspects of the theory (although one could, with justification, argue that the fussy details are at the physical heart of the problem). So we will limit considerations to a one-dimensional scalar field $\phi(x, t)$, which depends only on a single spatial coordinate x and on time t, rather than a three-dimensional pair of fields required for real electromagnetism. The transition amplitude that describes the field's behavior is, according to the equation for the KEY-PATH INTEGRAL,

$$K(\phi(x,t), \phi(\xi,\tau)) = \int \mathcal{D}\phi \exp\left(\frac{i}{\hbar}\int\int dxdt\, \mathcal{L}(\phi, \partial_\mu \phi)\right).$$

Note that the argument in the exponential function is now an integral over both time and space and not just time, as it was earlier. The Lagrangian L of the earlier treatment has been replaced by the Lagrangian density $\mathcal{L}$—this is the field's Lagrangian—which must be integrated over the spatial coordinates; why this is so will be explained by example in the next section.

For the electromagnetic field, the $\mathscr{L}$ that appears in the preceding equation is the one associated with Maxwell's equations. We will use a simpler one: expand $\mathscr{L}(\phi, \partial_\mu \phi)$ in a power series to lowest order in $\partial_\mu \phi$ (which is second order in this case, to keep the symmetry of the Lagrangian density in time and space) to get

$$\mathscr{L}(\phi, \partial_\mu \phi) = (\partial_\mu \phi)(\partial_\mu \phi).$$

Here $\partial_\mu = (\partial_x, (i/c)\partial_t)$, with $c =$ the speed of light, and the repeated indices μ indicate the Einstein summation convention (page 55). This Lagrangian is the simplest possible one that is covariant and treats space as symmetric (which spatial coordinate is x and which is $-x$ are matters of convention rather than physical significance). As we will see in the next section, it will describe the "electromagnetic" field with a simple wave equation. The calculations will use this Lagrangian density, but I will frequently talk as if this Lagrangian is the one that gives Maxwell's equations; none of the essential points to be made depends on the precise form of the Lagrangian I use.

The Lagrangian for a Mechanical Field

To extend the Feynman formulation of quantum mechanics to a field, one must have an expression for the field's Lagrangian. The point of this section is just to show how such a Lagrangian for a field arises for a simple mechanical case.

Picture a linear chain of identical, massless, ideal springs connecting N point masses, each with mass $m = M/N$ (M is the total mass). These masses are assumed to move only in one dimension (x), so the springs are either stretched or compressed. The chain extends from $x = 0$ to $x = X$ (the ends of the chain cannot move), and the deviation of the kth mass point from its rest position is ϕ_k. When all of the ϕ_k are 0 (no displacements), the mass points are evenly spaced. For this system, the Lagrangian L is just kinetic energy minus potential energy:

$$L = \sum_{k=1}^{N} \frac{m\dot{\phi}_k^2}{2} - \sum_{k=0}^{N} V_k,$$

where V_k is the potential energy of the kth spring and $\dot{\phi}_k$ is the time derivative of the kth displacement. Hooke's law, with spring constant K, applied to the kth spring says that $V_k = (K/2)\,(\phi_{k+1} - \phi_k)^2$, so the Lagrangian is

$$L = \sum_{k=1}^{N} \frac{m\dot{\phi}_k^2}{2} - \frac{K}{2} \sum_{k=0}^{N} (\phi_{k+1} - \phi_k)^2.$$

Now find the limit as the number of springs $N \to \infty$ in such a way that

$$\Delta x = \frac{X}{N+1}$$

$$m = \frac{M}{N}$$

$$K\Delta x = K_0,$$

with X, M, and K_0 fixed. K_0 is the spring constant for the entire chain of springs. Define the mass per unit length as $\mu = M/X$. Thus

$$\mu = \frac{M}{X} = \frac{Nm}{(N+1)\Delta x},$$

so

$$m = \frac{N+1}{N}\mu\Delta x \quad \text{and} \quad K = \frac{K_0}{\Delta x}.$$

Using these relationships, write the Lagrangian as

$$L = \frac{1}{2}\sum_{k=1}^{N} \dot{\phi}_k^2 \underbrace{\frac{N+1}{N}\mu\Delta x}_{m} - \frac{1}{2}\sum_{k=0}^{N} K_0 \left(\frac{\phi_{k+1} - \phi_k}{\Delta x}\right)^2 \Delta x.$$

As $N \to \infty$,

$$L \to \frac{1}{2}\int_0^X \left(\mu\left(\frac{\partial\phi}{\partial t}\right)^2 - K_0\left(\frac{\partial\phi}{\partial x}\right)^2\right) dx.$$

The Lagrangian density $\mathscr{L}$ can thus be identified as

$$\mathscr{L} = \mu\left(\frac{\partial\phi}{\partial t}\right)^2 - K_0\left(\frac{\partial\phi}{\partial x}\right)^2.$$

Note that the action for this mechanical field ϕ is

$$S[\phi(x,t)] = \int L\,dt = \int\int dx dt\, \mathscr{L}(\dot{\phi}),$$

where $\dot{\phi}$ as an argument of $\mathscr{L}$ indicates both the temporal and spatial derivatives of ϕ.

Now we wish to identify the Euler-Lagrange equation that corresponds to this Lagrangian. That it will turn out to be a wave equation may come as no surprise. In anticipation of our later work, select the constants in the Lagrangian density $\mathscr{L}$ so that it now reads

$$\mathscr{L}(\dot{\phi}) = \frac{1}{c^2}\left(\frac{\partial\phi}{\partial t}\right)^2 - \left(\frac{\partial\phi}{\partial x}\right)^2.$$

The EULER-LAGRANGE EQUATION is

$$\frac{\partial\mathscr{L}}{\partial\phi} - \frac{\partial}{\partial t}\left(\frac{\partial\mathscr{L}}{\partial_t\phi}\right) - \frac{\partial}{\partial x}\left(\frac{\partial\mathscr{L}}{\partial_x\phi}\right) = 0.$$

So the Lagrangian density $\mathscr{L}(\dot{\phi})$ gives, as its Euler-Lagrange equation, the *two-dimensional wave equation*:

WAVE EQUATION
(TWO-DIMENSIONAL)

$$-\frac{1}{c^2}\frac{\partial^2\phi}{\partial t^2} + \frac{\partial^2\phi}{\partial x^2} = 0$$

Thus one can see how the Lagrangian density arises and why the integral that is the argument of the exponential function in the KEY-PATH INTEGRAL must include both space and time when we treat this system quantum-mechanically. Note that the Lagrangian density that arises from coupled harmonic oscillators, as described above, is the same as the one for the example below. Of course, the field we picture need not have this mechanical analog.

The Field-Transition Amplitude

Just as with ordinary quantum mechanics, the key quantity for field theory is the transition amplitude:

$$K(\phi(x,t),\phi(x',t')) = \int \mathscr{D}\phi \exp\left(\frac{i}{\hbar}\int_{\xi}^{x}\int_{\tau}^{t} dxdt\,\mathscr{L}(\partial_\mu\phi)\right)$$
$$= \int \mathscr{D}\phi \exp\left(\frac{i}{\hbar}\int d\xi\,\mathscr{L}\right)$$

In the last equation of this chain we have used the vector $\xi = (x,t)$, which, in the general case, would be four-dimensional. The integral in the argument of the exponential function uses the two-dimension volume element $d\xi \equiv d\xi_1 d\xi_2 = dxdt$; this convention will be used in the following discussion. According to the last equation, the field amplitude is determined probabilistically. Of course, in experiments we would measure only average values of the field amplitude.

The transition amplitude just given must be calculated anew for each starting and ending state. An alternative and more general approach is to start and end in a standard "blank" state and then insert the initial state with a driving function. The starting and ending states are taken to be the lowest-energy, *vacuum* states of the system, and we calculate the amplitude of the vacuum-to-vacuum transition. These states are often written $\langle 0|$ and $|0\rangle$, so we want the amplitude of the transition $\langle 0|0\rangle$. The interesting states are then produced by the driving function J (as I will describe below), and the resulting amplitude of the vacuum-to-vacuum transition with driving in between is called a *generating functional.* This generating functional just is like the one used, for the same reasons, to characterize the harmonic oscillator in chapter 4 (see page 131). Thus we start with the generating functional:

GENERATING FUNCTIONAL

$$W[J] = \int \mathscr{D}\phi \exp\left(\frac{i}{\hbar}\int_{-\infty}^{\infty} [\mathscr{L}(\phi,\partial_\mu\phi) + J\phi]\,d\xi\right)$$

Here $J(\xi)$ is a forcing function used to give some desired states. How this forcing works was illustrated on page 132. The system is assumed to start (at $t = -\infty$) and finish (at $t = \infty$) in a lowest-energy vacuum state, so $W[J]$ is specifically the amplitude of the vacuum-to-vacuum transition $W[J] = \langle 0|0\rangle$. The trick behind the generating functional is much like that used in probability theory for generating functions that produce all of the moments of a probability distribution. It will turn out that this generating functional can be used to calculate all of the specific transition amplitudes that we should want. How we use the generating functional is described in detail below.

The central entity for relating experimentally observable quantities to field-theoretic predictions is the *Green's function* for the field. Many experiments measure particle lifetimes and scattering cross sections, and these can be found from the Green's function (although just how this is done requires some lengthy calculations). Our goal now is to define the Green's function and relate it to the generating functional above. We will see that the Green's function is also identified with another function, the *Feynman propagator.* In turn, the propagator is closely associated with the Euler-Lagrange equation arising from the particle's Lagrangian. The explanation of what all of this means is the subject of this section and the next.

To recapitulate our current position, we have selected a Lagrangian for a field and have used the KEY-PATH INTEGRAL to compute the transition amplitude for this field. One can picture the electromagnetic field, but we are actually using a Lagrangian that gives a simpler two-dimensional scalar field. The Green's function for the field $\phi(x, t)$ is defined to be

$$G(\xi, \xi') \equiv \int \mathscr{D}\phi\, \phi(\xi)\phi(\xi') \exp\left(\frac{i}{\hbar}\int d\xi\, \mathscr{L}(\phi, \partial_\mu \phi)\right)$$

(remember that $d\xi = dxdt$). This gives, as before (see page 202), the probability of ending at $\phi(\xi)$, given that the system started at $\phi(\xi')$.

The Green's function is calculated from the generating functional $W[J]$ by functional differentiation according to the relation

$$G(\xi, \xi') = \left(\frac{\hbar}{i}\right)^2 \left.\frac{\delta^2 W[J]}{\delta J(\xi)\delta J(\xi')}\right|_{J=0}$$

because the function-differential operators bring down the functions ϕ that multiply J:

$$\frac{\delta^2 W[J]}{\delta J(\xi)\delta J(\xi')} = \frac{\delta^2}{\delta J(\xi)\delta J(\xi')} \int \mathscr{D}\phi \exp\left(\frac{i}{\hbar}\int_{-\infty}^{\infty} [\mathscr{L}(\phi, \partial_\mu \phi) + J\phi]\, d\xi\right)$$

$$= \left(\frac{i}{\hbar}\right)^2 \int \mathscr{D}\phi\, \phi(\xi)\, \phi(\xi') \exp\left(\frac{i}{\hbar}\int_{-\infty}^{\infty} [\mathscr{L}(\phi, \partial_\mu \phi) + J\phi]\, d\xi\right)$$

Those interested in analogies with probability theory should note that this is like the usual calculation for moments of a probability distribution from the probability-generating function and that G corresponds to the covariance.

There really is not just one Green's function but a whole family, whose members are denoted $G^{(n)}$; the nth member of this family is the nth-order functional derivative of $W[J]$ with respect to $J(\xi)$ evaluated at $J = 0$, and what I have called G is really $G^{(2)}$. Again, these family members are like the various moments calculated from a probability-generating function.

At this point a proper treatment of field theory would go in two directions. The first would be a development of the relationship between Green's functions and experimentally measurable quantities; we are going to take on faith that such a relationship exists. The second direction is an examination of the meaning of the Green's function and how this function might be explicitly calculated. We now embark on this course.

The Feynman Propagator

Using precisely the same steps presented on page 133 to simplify the quantum-mechanical GENERATING FUNCTIONAL, we may express $W[J]$ as a product of two factors: $W[0]$ and an exponential function that involves the driving function J. Specifically, the result is the *relation between the generating functional and the propagator*:

RELATION BETWEEN
THE GENERATING FUNCTIONAL AND THE PROPAGATOR

$$W[J] = W[0] \exp\left(\frac{i}{2\hbar} \int\int d\xi d\xi' \, J(\xi) \Delta_{\mathrm{F}}(\xi, \xi') J(\xi')\right)$$

Here $W[0]$ is the amplitude of the vacuum-to-vacuum transition in the absence of driving and is given by

$$W[0] = \int \mathscr{D}\phi \exp\left(\frac{i}{\hbar} \int d\xi \, \mathscr{L}\right);$$

the function $\Delta_{\mathrm{F}}(\xi, \xi')$, termed the *Feynman propagator*, is defined below. This expression is crucial because the generating functional, the object needed to relate the theory to experimental measurements, can be calculated from the constant $W[0] = 1$ (because the system starting in the vacuum state surely ends there without driving) and an expression that depends only on the driving function.

The Green's function G is seen to be just

$$G(\xi, \xi') = \left(\frac{i}{\hbar}\right)^2 \left[\frac{\delta^2 W[J]}{\delta J(\xi)\delta J(\xi')}\right]_{J=0}$$

$$= \left(\frac{i}{\hbar}\right)^2 \left[\frac{\delta^2}{\delta J(\xi)\delta J(\xi')} \exp\left(\frac{i}{\hbar}\int\int d\xi d\xi' \, J(\xi)\Delta_F(\xi, \xi')J(\xi')\right)\right]_{J=0}$$

$$= \frac{\hbar}{i}\Delta_F(\xi, \xi').$$

The Feynman propagator Δ_F is defined to be the negative of the impulse response of the EULER-LAGRANGE EQUATION that corresponds to the Lagrangian $\mathscr{L}$ (look back at chapter 4, page 134, where the function there that corresponds to Δ_F is called D).

The Lagrangian for the "electromagnetic" field gives the WAVE EQUATION as its related EULER-LAGRANGE EQUATION. What is the propagator associated with this wave equation, and what is its physical interpretation? To find the impulse response for the wave-equation operator, start by applying the Fourier transform, defined as

$$\mathscr{F}[\Delta_F(x, t)] = \int_{-\infty}^{\infty} dx \int_{-\infty}^{\infty} dt \, \Delta_F(x, t) e^{-i(\omega t + xp/\hbar)} \equiv \hat{\Delta}_F(\omega, p),$$

to the wave equation with δ-function driving:

$$\frac{1}{c^2}\frac{\partial^2 \Delta_F}{\partial t^2} - \frac{\partial^2 \Delta_F}{\partial x^2} = \delta(x)\delta(t),$$

where $\Delta_F(x, t)$ is the required impulse response (propagator). Remember that when moving from position space to momentum space, the transform variable that corresponds to x must be $p/\hbar$ (see page 115 in chapter 4). The Fourier-transformed Green's function $\hat{\Delta}_F(\omega, p)$ depends on frequency ω and momentum p. Because $\mathscr{F}[\partial^2\Delta_F/\partial t^2] = -\omega^2\hat{\Delta}_F$ and $\mathscr{F}[\partial^2\Delta_F/\partial x^2] = -(p/\hbar)^2\hat{\Delta}_F$, the equation for $\hat{\Delta}_F$ becomes

$$\left(-\frac{\omega^2}{c^2} + \frac{p^2}{\hbar^2}\right)\hat{\Delta}_F(\omega, p) = \left(\frac{-\omega^2\hbar^2 + p^2c^2}{c^2\hbar^2}\right)\hat{\Delta}_F = 1,$$

or (when you solve for $\hat{\Delta}_F$)

$$\hat{\Delta}_F(\omega, p) = \frac{-c^2\hbar^2}{\omega^2\hbar^2 - p^2c^2} = \frac{-c^2\hbar^2}{(\omega\hbar - \rho c)(\omega\hbar + pc)}.$$

This is the impulse response for the wave-equation operator in momentum/frequency space. The propagator in position/time space is given by the inverse Fourier transform

$$\Delta_F(x,t) = \frac{1}{(2\pi)^2\hbar}\int_{-\infty}^{\infty} dp\, e^{ixp/\hbar}\int_{-\infty}^{\infty} d\omega\, e^{i\omega t}\frac{-c^2\hbar^2}{(\omega\hbar - cp)(\omega\hbar + cp)}$$

$$= \frac{c^2\hbar}{(2\pi)^2}\int_{-\infty}^{\infty} dp\, e^{ixp/\hbar}\int_{-\infty}^{\infty} d\omega\, e^{i\omega t}\frac{-1}{(\omega\hbar - cp)(\omega\hbar + cp)}.$$

To perform this inverse Fourier transform, look first at the integral over ω:

$$\frac{1}{2\pi}\int_{-\infty}^{\infty} d\omega\, e^{i\omega t}\frac{1}{(\omega\hbar - cp)(\omega\hbar + cp)}$$

This integration requires that the poles on the ω axis be properly handled; the treatment of these poles is equivalent to deciding on the boundary conditions for the equation. The idea is to place the poles just above or just below the ω axis so that they make a contribution specified by the residue theorem when a closed integration path that goes through $\pm i\infty$ is selected. This is a standard procedure for inserting initial conditions for this situation, and the final solution depends on just how the poles are treated. When the pole is displaced slightly from the ω axis and the Fourier transform is changed into a contour integral by extending ω into the complex plane, we have

$$\frac{1}{2\pi}\int_{-\infty}^{\infty} d\omega\, e^{i\omega t}\frac{1}{(\omega\hbar - cp)(\omega\hbar + cp)} = \frac{1}{2\pi}\oint d\omega\, e^{i\omega t}\frac{1}{(\omega\hbar - cp)(\omega\hbar + cp)}$$

because the semicircle in the imaginary plane at infinity contributes nothing to the integral if we close the contour in the proper half plane. For example, if the argument of the exponential is negative, we close the contour through $+i\infty$. We shall return to this question of poles later (page 220), but for the time being, we place the positive pole (the one at $\omega\hbar$) on the negative side of the ω axis and the negative pole on the positive side of the ω axis. When the path for $t > 0$ is closed through $i\infty$, only the pole at $-\omega\hbar$ contributes ($\omega = -cp/\hbar$), so the residue is $-2\pi i e^{-tcp/\hbar}/2cp$ and the integral becomes

$$\frac{1}{2\pi}\oint d\omega\, e^{i\omega t}\frac{1}{(\omega\hbar - cp)(\omega\hbar + cp)} = -\frac{i}{2cp\hbar}e^{-icpt/\hbar}.$$

Now use this result to perform the Fourier transform over momentum for $t > 0$ to give the required impulse response:

$$\Delta_F(x,t) = \frac{ic}{4\pi}\int_{-\infty}^{\infty} dp \frac{e^{i(x-ct)p/\hbar}}{p} = \frac{-c\hbar}{2}\theta(x-ct),$$

where $\theta(y)$ is the unit step function ($\theta(y) = 1$ for $y > 0$ and $\theta(y) = 0$ otherwise). To see how the step function arises, define

$$\theta(y) \equiv \frac{-i}{2\pi}\int_{-\infty}^{\infty} dk \frac{e^{iky}}{k},$$

and look at the derivative of this function:

$$\frac{d\theta}{dy} = \frac{d}{dy}\left(\frac{-i}{2\pi}\int_{-\infty}^{\infty} dk \frac{e^{iky}}{k}\right) = \frac{1}{2\pi}\int_{-\infty}^{\infty} dk\, e^{iky} = \delta(y)$$

(See page 23 for the definition of the DIRAC DELTA FUNCTION and page 122 for its Fourier representation.) Thus $\theta(y)$ is the unit step function because its derivative is a delta function.

The "electromagnetic"-field propagator is therefore

$$\Delta_F(x,t;x',t') = \frac{-c\hbar}{2}\theta((x-x') - c(t-t')).$$

This event, i.e., the traveling edge of the step function, moves from x' at t' to x at t with the velocity of light c. Furthermore, the energy associated with the transmitted wave is $E = \omega\hbar$.

To see why this is true, we need to find the momentum of the wave: the energy E of a massless, free-moving particle is given by $E = cp$, a relation that comes from the expression $E^2 = c^2p^2 + m^2c^4$ (page 192), with $m = 0$. Note that the traveling wave must have no mass, because it moves at the speed of light and a massive object cannot move with that velocity. We started with an expression for the Green's function in momentum/frequency space and used inverse Fourier transforms to find the position/time version. The original momentum/frequency equation contained poles that determined the initial conditions of the wave, and the positive energy pole dictates that $\hbar\omega = cp$. Therefore, our wave constitutes an object that propagates and has a quantized energy of $E = \hbar\omega$ (often ν rather than ω is used to denote the frequency of the electromagnetic field, so

$E = \hbar\nu$). Because the propagator describes the movement at the speed of light of an object with energy proportional to its frequency (the proportionality constant is $\hbar$), we identify this object as the "photon" of the scalar "electromagnetic" field.

Let us take stock. We started with the Lagrangian for a classical "electromagnetic" field and used the KEY-PATH INTEGRAL (page 202) to find the transition amplitude for an "electromagnetic" field going from the lowest-energy vacuum state at $t = -\infty$ to the same state after a long time ($t = \infty$). We calculated this transition amplitude just as we did for a particle when going from classical to quantum mechanics in chapter 4. This transition amplitude turned out to provide us with the Green's function (or equivalently, the propagator) that is the amplitude for going from one field state to another. When the propagator is found, a quantum of energy that propagates with the speed of light, a "photon," emerges. Thus the quantal nature of the electromagnetic field arises naturally from the same sort of process that earlier led to the quantum-mechanical nature of particles.

The next step is to combine this quantum mechanics of the field with that of particles, as described in chapter 4. Before taking this step, however, we must further examine the significance of the propagator.

The fourth-order Green's function is gotten from the generating functional as described earlier:

$$G(\xi_1, \xi_2; \xi_1', \xi_2') = \left(\frac{\hbar}{i}\right)^4 \left[\frac{\delta^4 W[J]}{\delta J(\xi_1)\delta J(\xi_2)\delta J(\xi_1')\delta J(\xi_2')}\right]_{J=0}$$

If the system consists of particles, this function specifies the amplitude for particle 1 to go from ξ_1' to ξ_1 and for particle 2 to go from ξ_2' to ξ_2. The sixth-order Green's function would, in turn, describe the amplitude for three particles, and so on. Thus the generating functional describes the movement of any number of particles whenever the Lagrangian can be given a particle interpretation.

Because of the RELATION BETWEEN THE GENERATING FUNCTIONAL AND THE PROPAGATOR, displayed above, all of the various orders of Green's functions can be calculated just from the propagator Δ_F. Thus the entire description of a quantum system with any number of particles is in hand once the propagator is known.

Second Quantization

Recall that the formulation of quantum mechanics presented in chapter 4 suffered from two main problems. The first was that Schrödinger's equation is not covariant and a relativistic quantum mechanics is required for rapidly moving particles; the second was that the formulation did not easily lend itself to describing systems in which the number of particles is changing, as they do in many microphysical situations. The solution to the first problem is to develop a covariant Lagrangian to replace the one that gives rise to the Schrödinger equation. For simplicity, we will avoid going in this direction because the main characteristics of a covariant theory are, for our purposes, illustrated by the preceding treatment of the "electromagnetic" field in which the Lagrangian is covariant, as I will discuss below. We thus turn to the second problem, the difficulty of describing systems in which the particle number is changing as the system evolves. As an example of the general approach, we will extend the traditional nonrelativistic single-particle quantum mechanics, the quantum mechanics of chapter 4, in a way that naturally includes a description with various numbers of particles.

The generating functional used above naturally dealt with this problem of varying numbers of particles because the various orders of Green's functions, all of which can be calculated from just the propagator, describe systems with different numbers of particles. The question, then, is how to find the Lagrangian appropriate for describing traditional quantum mechanics in the framework used above for the "electromagnetic" field. The key is this observation: the propagator calculated from the Euler-Lagrange equation describes the amplitude for a particle to go from x', t' to x, t, and this is what the transition amplitude $K(x, t; x', t')$ used in chapter 4 does. Further, $K(x, t; x', t')$ is just the Green's function for the Schrödinger equation (see page 112). If we find the Lagrangian that has the Schrödinger equation as its Euler-Lagrange equation—we denote this Lagrangian as $\mathscr{L}_S$—and use this Lagrangian in the KEY-PATH INTEGRAL, just as we did the one for the "electromagnetic" field above, the resulting formulation will express the system's behavior in terms of the propagator that is the transition amplitude of chapter 4. That is, the propagator for this situation reproduces the single-particle theory of chapter 4. But now the inclusion of different particle numbers becomes easy because all of the various orders of Green's functions, which describe systems with different numbers of particles, can be directly calculated from just the propagator, as above. Thus the sequence is

$\mathcal{L}_S \rightarrow$ Schrödinger (Euler-Lagrange) equation

$\rightarrow$ propagator = single-particle transition amplitude

$\rightarrow$ higher-order Green's functions = description of multiparticle case.

Finding a Lagrangian that gives Schrödinger's equation as its Euler-Lagrange equation is not difficult. This Lagrangian must depend on $\psi(x, t)$ and on the derivatives of this function. The complex conjugate of Schrödinger's equation in which ψ^* replaces ψ is, however, an equally valid description of the physical situation because the probabilities of various events are calculated from $\psi^*\psi$. This means that ψ^* must appear symmetrically in the Lagrangian so that Schrödinger's equation or its complex conjugate could equally arise.

Recall that SCHRÖDINGER'S EQUATION for the world with a single spatial dimension is

$$\underbrace{-\frac{\hbar}{i}\frac{\partial\psi}{\partial t}}_{(3)} = \underbrace{-\frac{\hbar^2}{2m}\frac{\partial^2\psi}{\partial x^2}}_{(2)} + \underbrace{V(x)\psi}_{(1)}$$

(see page 111). Thus in the Lagrangian density $\mathcal{L}_S$ we need terms that include ψ and ψ^* in a symmetric way and that, through the general EULER-LAGRANGE EQUATION,

$$\underbrace{\frac{\partial\mathcal{L}}{\partial\psi}}_{(1)} - \underbrace{\frac{\partial}{\partial x}\left(\frac{\partial\mathcal{L}}{\partial(\partial_x\psi)}\right)}_{(2)} - \underbrace{\frac{\partial}{\partial t}\left(\frac{\partial\mathcal{L}}{\partial(\partial_t\psi)}\right)}_{(3)} = 0,$$

will yield each of the terms in Schrödinger's equation. We find the required Lagrangian by examining three terms that, it will be apparent, are the right ones to use for construction the Lagrangian.

First, suppose that $\mathcal{L}_S$ contains a term like $V\psi^*\psi$, and operate on this term with $\partial/\partial\psi^*$:

$$\frac{\partial}{\partial\psi^*}V\psi^*\psi = V(x)\psi$$

This operation gives the potential-energy term in Schrödinger's equation. Second, consider

$$\frac{\partial}{\partial x}\left(\frac{\partial}{\partial(\partial_x\psi^*)}\left[\frac{\hbar^2}{2m}(\partial_x\psi^*\partial_x\psi)\right]\right) = \frac{\hbar^2}{2m}\frac{\partial^2\psi}{\partial x^2},$$

which gives the spatial term in Schrödinger's equation. Third and finally, we have

$$\frac{\partial}{\partial t}\left(\frac{\partial}{\partial(\partial_t\psi^*)}\left[\frac{i}{\hbar}(\psi^*\partial_t\psi - \psi\partial_t\psi^*)\right]\right) = \frac{i}{\hbar}\frac{\partial\psi}{\partial t}$$

for the temporal term in Schrödinger's equation. Thus the Lagrangian density

$$\mathscr{L}_S \equiv \underbrace{-\frac{i}{\hbar}(\psi^*\partial_t\psi - \psi\partial_t\psi^*)}_{(3)} - \underbrace{\frac{\hbar^2}{2m}(\partial_x\psi^*\partial_x\psi)}_{(2)} + \underbrace{V\psi^*\psi}_{(1)}$$

has Schrödinger's equation (and its complex conjugate, found by using ψ rather than ψ^* in taking derivatives of $\mathscr{L}_S$) as its Euler-Lagrange equation. The GENERATING FUNCTIONAL,

$$W[J, J^*] = \int\int \mathscr{D}\psi\mathscr{D}\psi^* \exp\left(\frac{i}{\hbar}\int\int [\mathscr{L}_S + J^*\psi + J\psi^*]\,dxdt\right),$$

is associated with the propagator $\Delta_F(x', t'; x, t) = K(x, t; x', t')$ used to describe the motion of a single particle quantum-mechanically in chapter 4, and this propagator is just the solution for Schrödinger's equation with delta-function driving. Systems with various particle numbers are described by constructing the appropriate-order Green's functions from this propagator, as indicated above.

The procedure just described is usually called *second quantization*. First quantization, which was the subject of chapter 4, provides a quantum-mechanical description of a single particle's behavior by using the classical particle Lagrangian in the KEY-PATH INTEGRAL. The result of this is a field $\psi(x, t)$ that specifies the particle behavior probabilistically. For second quantization, a particular Lagrangian $\mathscr{L}_S$ is found that replaces the position and its derivatives in the Lagrangian with the first-quantized field ψ and its derivatives. Now the first-quantized result is contained within the second-quantized formulation because the second-quantized propagator is identical to the first-quantized transition amplitude. But this new formulation extends the theory to systems with variable numbers of particles. All of this worked because of the particular interrelation of Lagrangians, Euler-Lagrange equations, Green's functions of various orders, and the propagator that appears in the generating functional.

This same result can be gotten another way. In fact, it was historically obtained by a different route before path integrals came into use. The first quantization promotes classical observables, like position and momentum, to the status of operators that act on Hilbert-space vectors. The second quantization analogously promotes wave functions to operator status. This can be seen by comparing the variables on which the Lagrangians depend in the two cases: whereas x and its derivatives appear in the first-quantized Lagrangian, ψ and its derivatives appear in the second-quantized Lagrangian. The alternative path to second quantization thus makes wave functions into operators and uses the OPERATOR EQUATION OF MOTION for the Heisenberg picture as the basis for the description.

Interacting Fields

The preceding discussion has focused on free fields, but, of course, the most interesting problems relate to interactions. This section sketches how interactions are included in the theory.

The basic idea is very easy: replace the free-field Lagrangian in the KEY-PATH INTEGRAL with one that includes the interactions. For example, the "electromagnetic"-field Lagrangian is

$$\mathscr{L}(\dot{\phi}) = \frac{1}{c^2}(\partial_t\phi)^2 - (\partial_x\phi)^2,$$

and the Schrödinger Lagrangian is

$$\mathscr{L}_{\mathrm{S}} = \frac{i}{\hbar}(\psi^*\partial_t\psi - \psi\partial_t\psi^*) - \frac{\hbar^2}{2m}(\partial_x\psi^*\partial_x\psi) + V\psi^*\psi,$$

so the total Lagrangian $\mathscr{L}_{\mathrm{T}}$, including interactions, can always be written as something like

$$\mathscr{L}_{\mathrm{T}} = \mathscr{L} + \mathscr{L}_{\mathrm{S}} + \underbrace{\lambda\psi^*\phi\psi}_{\text{interaction}}.$$

Here the Lagrangian density for the interaction has been taken as having the very simple form $\lambda\psi^*\phi\psi$, where λ is the *coupling constant* that characterizes the strength of the interaction. Clearly, this interaction term couples the two fields, because changes in one will influence the other, since the interaction term alters the action. In general, of course, the interaction Lagrangian might be much more complicated, and finding the right one is where the physics is. As before, we can define a generating functional for the interacting system that gives the amplitude, under forcing functions J, J^*, and F, for the vacuum-to-vacuum transition:

$$W[J, J^*, F] = \int\int\int \mathscr{D}\psi\mathscr{D}\psi^*\mathscr{D}\phi \times \exp\left(\frac{i}{\hbar}\int\int dxdt\,(\mathscr{L} + \mathscr{L}_S + \lambda\psi^*\phi\psi + J\psi^* + J^*\psi + F\phi)\right)$$

Again as before, the Green's functions of all orders are calculated by taking the functional derivatives of W with respect to the forcing functions; these Green's functions describe the system.

The problem is that the interaction term ruins the simplicity of the free-particle results. The RELATION BETWEEN THE GENERATING FUNCTIONAL AND THE PROPAGATOR no longer holds for this situation, so the Green's functions can no longer be calculated in terms of the free-field propagators. The trick is to reformulate the equation above so that we can make use of the free-field propagators; that is, we want to express W in terms of the free-field case. To do this, define the free-field generating functional as

$$W_0[J, J^*, F] = \int\int\int \mathscr{D}\psi\mathscr{D}\psi^*\mathscr{D}\phi \times \exp\left(\frac{i}{\hbar}\int\int dxdt\,(\mathscr{L} + \mathscr{L}_S + J\psi^* + J^*\psi + F\phi)\right).$$

This is just the generating functional without the interaction term, so the fields all behave independently. Initially, assume that λ is so small that the field interaction can be approximated by

$$\exp\left(\frac{i}{\hbar}\int\int dxdt\,\lambda\psi^*\phi\psi\right) \approx 1 + \frac{\lambda i}{\hbar}\int\int dxdt\,\psi^*\phi\psi.$$

This means that W can be written approximately as

$$\iiint \mathscr{D}\psi\mathscr{D}\psi^*\mathscr{D}\phi \exp\left(\frac{i}{\hbar}\iint dxdt\,(\mathscr{L} + \mathscr{L}_S + J\psi^* + J^*\psi + F\phi)\right)$$

$$\times \underbrace{\exp\left(\frac{\lambda i}{\hbar}\iint dxdt\,\psi^*\phi\psi\right)}_{\text{interaction}}$$

$$\approx \iiint \mathscr{D}\psi\mathscr{D}\psi^*\mathscr{D}\phi \exp\left(\frac{i}{\hbar}\iint dx\,dt\,(\mathscr{L} + \mathscr{L}_S + J\psi^* + J^*\psi + F\phi)\right)$$

$$\times \left(1 + \frac{\lambda i}{\hbar}\iint dx\,dt\,\psi^*\phi\psi\right)$$

$$= W_0 + \frac{\lambda i}{\hbar}\iiint \mathscr{D}\psi\mathscr{D}\psi^*\mathscr{D}\phi$$

$$\times \exp\left(\frac{i}{\hbar}\iint dx\,dt\,(\mathscr{L} + \mathscr{L}_S + J\psi^* + J^*\psi + F\phi)\right)$$

$$\times \underbrace{\left(\iint dx\,dt\,\psi^*\phi\psi\right)}_{\text{interaction}}.$$

The result of this approximation is to bring down the interaction integral from the exponential function. This same end can be achieved by applying to W_0 the operator $\lambda\mathbf{R}$, defined as

$$\lambda\mathbf{R} \equiv \lambda\left(\frac{\hbar}{i}\right)^2 \iint dxdt \frac{\delta}{\delta J^*(x,t)}\frac{\delta}{\delta J(x,t)}\frac{\delta}{\delta F(x,t)}.$$

The result of $\lambda\mathbf{R}W_0$ is to bring down the various fields from the exponential function and integrate them over their spatial and temporal variables. The trick is that applying $\mathbf{R}$ in effect brings the interaction Lagrangian outside of the functional integral, so that we are left with operations on W_0. But according to the relation between the generating functional and the propagator, the free-field generating functional can be expressed in terms of the field propagators, just as before. Thus interactions are calculated, to this approximation, as functions that we know. For example, the approximation of the second-order Green's function is found from

$$G(x,t;x',t') \approx \frac{\delta^2}{\delta J^* \delta J^*} \frac{\delta^2}{\delta J \delta J} \frac{\delta^2}{\delta F \delta F} (W_0 + \lambda \mathbf{R} W_0) \Big|_{J,J^*,F=0},$$

where each functional derivative is evaluated at (x, t) and (x', t'); that is,

$$\delta^2/\delta F \delta F \equiv \delta^2/\delta F(x,t) \delta F(x',t'),$$

etc. This Green's function includes the interactions to the extent that the first-order approximation of the exponential function

$$\exp\left(\frac{i}{\hbar} \int\int dxdt \, \lambda \psi^* \phi \psi\right)$$

is adequate. Higher-order approximations can be found in the same way. For example, to include a second-order term, the expansion of the exponential function containing the interaction Lagrangian would be

$$\exp\left(\frac{i}{\hbar} \int\int dxdt \, \lambda \psi^* \phi \psi\right)$$

$$\approx 1 + \frac{\lambda i}{\hbar} \int\int dxdt \, \psi^* \phi \psi + \lambda^2 \left(\frac{i}{2\hbar} \int\int dxdt \, \psi^* \phi \psi\right),$$

and the same result can be achieved by applying the operator $1 + \lambda \mathbf{R} + (\lambda^2/2)\mathbf{R}^2$ to W_0. In general,

$$\exp\left(\frac{i}{\hbar} \int\int dxdt \, \lambda \psi^* \phi \psi\right) \Leftrightarrow e^{\lambda \mathbf{R}}$$

in the sense that applying the first k terms in the expansion of the $e^{\lambda \mathbf{R}}$ operator gives just the same result as expanding the interaction term within the functional integral to k terms. Although messy expressions can arise as higher-order terms are included, the problem of interactions is in principle solved in terms of free-field propagators that we know. For electrodynamics (with a real electromagnetic field, and a relativistic electron described by the Dirac equation), $\lambda = 1/137$, so only a few terms in the expansion give a very accurate approximation.

Antiparticles

In the discussion of the "electromagnetic" field, time and space appeared symmetrically in the Lagrangian, so the Lagrangian is covariant. This symmetry produced two poles in the momentum/frequency-space representation of the propagator $\hat{G}(\omega, p)$. However, the space/time propagator $\Delta_F(x, t)$ we calculated (page 210) arose from only one of the two poles that resulted when we Fourier-transformed the wave equation. What about the neglected pole? The solution for the propagator used above corresponded to $t > 0$, and the inverse-Fourier-transform integral was carried out by closing the contour in the upper half plane. The $t < 0$ solution is the same except that it corresponds to a negative-energy "photon" that propagates backward in time. Because of the requirement that the Lagrangian be covariant, satisfactory field theories always exhibit these negative-energy solutions as well as the expected positive-energy ones. The negative-energy solutions cannot be simply discarded, as we did above. In fact, it would be undesirable to do so, because they correspond to *antiparticles*. A negative-energy particle propagating backward in time was interpreted (first by Feynman) as a positive-energy antiparticle propagating forward in time. The prediction of an antiparticle, the positron, was one of the great triumphs of the Dirac relativistic theory of the electron. Because of the symmetry introduced by the requirement that Lagrangians for fields be relativistically covariant, all particles have their antiparticle counterparts.

This presentation of field theory has been unsatisfactory in at least two ways. First, the relationship between the Green's functions and quantities that can be measured in the laboratory—scattering cross sections and particle lifetimes—has not been developed. This relationship is known as the *reduction formulas*, and it permits experimentally measured quantities to be found in terms of the various orders of Green's function, which in turn are expressed as combinations of the free-field propagators. Second, real Lagrangians have not been used, so real particles have not resulted from our calculations. The construction of satisfactory Lagrangians is at the core of field theory, and it turns on exploiting various *symmetries*. A symmetry is present in a Lagrangian when the Lagrangian is invariant under a particular transformation. For example, a Lagrangian whose form is unchanged by a Lorentz transformation is said to exhibit the symmetry required by relativity. (This is like saying that a square exhibits a symmetry when what you see is invariant under the transformation of rotating by 90°.) Other symmetry requirements are also placed on satisfactory

Lagrangians. The result of the full theory is a beautiful account of the forces of nature and the structure of the particles that these forces feel. Although electromagnetism and subatomic forces are very well explained by this scheme, the first classical field, gravity, has so far eluded inclusion. Current efforts aim at producing a grand unified theory that will account not only for electromagnetism and the weak and strong forces that are present at subatomic levels but also for gravity.

Additional Reading

Books are listed in the order of more general to more specialized.

1 Mathematics

Mathews, J., and Walker, R. L. *Mathematical Methods of Physics.* Second edition. W. A. Benjamin, Menlo Park, California, 1970. One of the standard surveys of the mathematical methods needed for theoretical physics. Much useful information presented in a usually clear and interesting way.

Schey, H. M. *Div, Grad, Curl, and All That: An Informal Text on Vector Calculus.* W. W. Norton and Co., New York, 1973. A very readable treatment of vector analysis, designed especially for students who are learning the material as preparation for studying electricity and magnetism.

Halmos, P. R. *Finite Dimensional Vector Spaces.* Princeton University Press, Princeton, 1958. A classic description of operators on finite-dimensional vector spaces. This is done with an eye toward Hilbert spaces.

Volterra, V. *Theory of Functionals and of Integral and Integro-differential Equations.* Dover Publications, New York, 1959. A reprint of the classic work on the theory of functionals. Still interesting and readable.

Feynman, R. P., and Hibbs, A. R. *Quantum Mechanics and Path Integrals.* McGraw-Hill Book Co., New York, 1965. A description of path integrals in the context of quantum mechanics. The development is interesting and inspiring but has a number of small errors and can be very frustrating. The section on Gaussian processes treated from a functional-integral point of view is one of the few easily available.

Zwillinger, D. *Handbook of Differential Equations.* Second edition. Academic Press, Boston, 1992. An extremely useful and unusually clear book that concisely presents very many different ways of handling the differential equations that appear in physical problems. Makes you feel as if you can deal with any differential equation.

2 Classical Mechanics

Goldstein, H. *Classical Mechanics.* Second edition. Addison-Wesley Publishing Co., Reading, Massachusetts, 1980. The classic treatment of classical mechanics that sets the standard for what graduate students should know about the subject. Some parts are done historically, and others in a very modern way with an eye toward applications to contemporary physics.

Joos, G., and Freeman, I. M. *Theoretical Physics.* Third edition. Dover Publications, New York, 1966. A good discussion of classical mechanics from a more traditional point of view. A book for those with a taste for nineteenth-century literature.

3 Electricity and Magnetism

Jackson, J. D. *Classical Electrodynamics.* Second edition. John Wiley and Sons, New York, 1975. The standard treatment of classical electricity and magnetism with emphasis on topics needed for other and more modern areas of physics.

Buchwald, J. Z. *From Maxwell to Microphysics: Aspects of Electromagnetic Theory in the Last Quarter of the Nineteenth Century.* University of Chicago Press, Chicago, 1985. A good history of an exciting era of physics. The history gives a good perspective on the theory.

Debye, P. *Polar Molecules.* Dover Publications, New York, 1929. A reprint of a classic work by one of the masters, this book provides wonderful insight into dielectric properties by studying molecules in a gas.

4 Quantum Mechanics

Sakurai, J. J. *Modern Quantum Mechanics.* Revised edition. Addison-Wesley Publishing Co., Reading, Massachusetts, 1994. Notable for its contemporary outlook, clarity, and physical insight, this standard work is used by many first-year graduate courses.

Feynman, R. P., and Hibbs, A. R. *Quantum Mechanics and Path Integrals.* McGraw-Hill Book Co., New York, 1965. The treatment of quantum mechanics on which chapter 4 was based. You can spend a day on half a page of this book at times, but all the effort is rewarded. One comes away from this book in awe of Feynman's originality.

Dirac, P. A. M. *The Principles of Quantum Mechanics.* Fourth edition. Clarendon Press, Oxford, 1958. This book, by one of the giants of physics, first appeared in 1930 and defined the modern view of the subject. Still easy to read.

Schiff, L. I. *Quantum Mechanics.* McGraw-Hill Book Co., New York, 1955. An old but clear and systematic treatment of quantum mechanics with an emphasis on physical insight.

White, R. L. *Basic Quantum Mechanics.* McGraw-Hill Book Co., New York, 1966. A brief, very clear treatment of standard elemantary quantum mechanics in which every step is spelled out. Easy to read.

5 Statistical Physics

Reif, F. *Fundamentals of Statistical and Thermal Physics.* McGraw-Hill Book Co., New York, 1965. A readable treatment of thermodynamics and statistical mechanics from a modern viewpoint. One of the standard texts.

Fermi, E. *Thermodynamics.* Dover Publications, New York, 1936. A brief, very clear, and systematic treatment of classical thermodynamics by a great physicist.

Callen, H. B. *Thermodynamics and an Introduction to Thermostatistics.* Second edition. John Wiley and Sons, New York, 1985. A very personal, unified treatment of thermodynamics that is a pedagogical classic.

Uhlenbeck, G. E., and Ford, G. W. *Lectures in Statistical Mechanics.* American Mathematical Society, Providence, Rhode Island, 1963. A brief, advanced treatment of classical statistical mechanics by one who loved the subject.

Brush, S.G. *Statistical Physics and the Atomic Theory of Matter from Boyle and Newton to Landau and Onsager.* Princeton University Press, Princeton, 1983. Recent years have seen an increased interest in the history of probability and statistics and their applications. This book provides a history of the growth of statistical ideas in physics.

6 Special Relativity

Goldstein, H. *Classical Mechanics.* Second edition. Addison-Wesley Publishing Co., Reading, Massachusetts, 1980. The best brief treatments of special relativity at the moderately advanced level appear either in the first part of books on general relativity or as part of books on classical mechanics or on electricity and magnetism. Goldstein's chapter on relativity is an especially good one.

Taylor, E. F., and Wheeler, J. A. *Spacetime Physics.* W. H. Freeman and Co., New York, 1966. A straightforward elementary introduction to special relativity by master teachers who really know the field.

7 Field Theory

Ramond, P. *Field Theory: A Modern Primer.* Second edition. Addison-Wesley Publishing Co., Redwood City, California, 1989. An up-to-date, modern, but fairly brief treatment that is widely read.

Aitchison, I. J. R., and Hey, A. J. G. *Gauge Theories in Particle Physics: A Practical Introduction.* Second edition. Adam Hilger, Bristol and Philadelphia, 1989. One of the clearer treatments of field theory, but without path integrals.

Bjorken, J. D., and Drell, S. D. *Relativistic Quantum Fields.* McGraw-Hill Publishing Co., New York, 1965. The old standard of field theory.

Itzykson, C., and Zuber, J.-B. *Quantum Field Theory.* McGraw-Hill Book Co., New York, 1980. A comprehensive, advanced treatment of the subject.

Hatfield, B. *Quantum Field Theory of Point Particles and Strings.* Addison-Wesley Publishing Co., Redwood City, California, 1992. A survey of standard field theory, as an introduction to string theory, that I found especially clear.

Symbol Index

Equation Index

Subject Index